AF181039

Edited by
Klaus Ammann
Yolande Jacot
Richard Braun

/ethods for Risk Assessment f Transgenic Plants

IV. Biodiversity and Biotechnology

Springer Basel AG

Editors:

Prof. Dr. Klaus Ammann
Dr. Yolande Jacot
Botanical Garden
Altenbergrain 21
CH – 3013 Bern
Switzerland

Prof. Dr. Richard Braun
BIOLINK
P.O. Box 208
CH – 3000 Bern 11
Switzerland

A CIP catalogue record for this book is available from the Library of Congress, Washington D.C., USA

Bibliographic information published by Die Deutsche Bibliothek
Die Deutsche Bibliothek lists this publication in the Deutsche Nationalbibliografie; detailed bibliographic data is available in the Internet at <http://dnb.ddb.de>.

ISBN 978-3-0348-9413-5 ISBN 978-3-0348-8033-6 (eBook)
DOI 10.1007/978-3-0348-8033-6

The publisher and editor can give no guarantee for the information on drug dosage and administration contained in this publication. The respective user must check its accuracy by consulting other sources of reference in each individual case.

The use of registered names, trademarks, etc. in this publication, even if not identified as such, does not imply that they are exempt from the relevant protective laws and regulations or free for general use.

This work is subject to copyright. All rights are reserved, whether the whole part of the material is concerned, specifically the rights of translation, reprinting, re-use of illustrations, recitation, broadcasting, reproduction on microfilms or in other ways, and storage in data banks. For any kind of use, permission of the copyright owner must be obtained.

© 2003 Springer Basel AG
Originally published by Birkhäuser Verlag in 2003

Computer-to-plate production, data prepared and provided by the editors
Printed on acid-free paper produced from chlorine-free pulp. TCF ∞

ISBN 978-3-0348-9413-5

9 8 7 6 5 4 3 2 1 www.birkhauser-science.com

Table of contents

Session 3: Impact of biotechnology on conservation

Session 4: Conclusions

Methods for Risk Assessment of Transgenic Plants
IV. Biodiversity and Biotechnology
K. Ammann, Y. Jacot and R. Braun (eds)
© 2003 Birkhäuser Verlag Basel/Switzerland

Introduction: Biodiversity – the impact of biotechnology

Richard Braun[1] and Klaus Ammann[2]

[1]*BIOLINK, P.O. Box 208, CH – 3000 Bern 11, Switzerland*
[2]*University of Bern, Botanical Garden, Altenbergrain 21, CH – 3013 Bern, Switzerland*

Drafted from a briefing paper of the European Federation of Biotechnology with editorial input of David Bennett.

Biodiversity is important for several very different reasons: the intrinsic value of species in the wild, the many varieties of plants, animals and micro-organisms used worldwide for farming and other human activities, as a genetic resource in healthcare, agriculture and food production, as well as for aesthetic and recreational purposes.

This paper reviews the various aspects of biodiversity in relation to biotechnology, including a chapter on gene flow in general. Biotechnology has the potential to improve sustainability in several ways and is expected, thereby, to help maintain natural as well as agricultural biodiversity. If adequately tested, products from GM crops may promote human health and a sustainable environment. Large monocultures may be problematic, irrespective of whether the crops are conventional or GM.

The essence of biodiversity

Biodiversity today is the result of 3.5 billions years of evolution. All the living organisms we know today, as well as those that ever lived before, have developed from one original micro-organism through the processes of mutation and selection. Separate species arose when mutations between relatives no longer allowed for interbreeding, for instance after geographic separation. A far greater variety of organisms once lived on our earth than live here today. The vast majority of species that arose, probably more than 99%, disappeared again. In the long-term view there has been no such thing as sustainability, only change. Clearly however today, with man's massive influence on the globe, change and the loss of biodiversity is much faster than at any time before, and therefore sustainability has to come into conservation and agricultural strategies, in order to avoid damage to life on earth.

The number of species of animals, plants and micro-organisms today is probably 10 million or more, of which only 1.4 million have yet been named. Virtually all the 40,000 vertebrate animals and most of the 250,000 higher plant species are known. On the other hand there are likely to be over a million species each of fungi and nematodes, and several million insects, of which only 70,000, 13,000 and 950000 respectively have been identified. Only about 5,000 bacteria and viruses have been identified individually, yet their total number may also be well in excess of one million.

The various species of plants and animals do not live an independent existence but are associated in specific communities and ecosystems to form more or less stable associations. One such association, for instance, is the tropical rain forest which is generally thought to have the highest degree of biodiversity. Other examples are specific types of alpine meadows or rivers and ponds. Streams, lakes and oceans are the habitats of a vast biodiversity of animals, plants and micro-organisms. Biodiversity needs to be considered both in the qualitative terms of the number of species present and also in the quantitative terms of how many individuals there are of each of the species. With large plants or animals this is quite easy to determine but impossible with micro-organisms. Often the number of species in a given ecosystem is taken as a measure of the biodiversity of that system because other criteria are more difficult to apply. In addition, it is still not clear whether the total amount of living matter at a given location, the biomass, is generally dependent on biodiversity: some researchers claim that loss of species is not compensated by additional growth of other species and thereby leads to a reduction in total biomass.

In agriculture, some 7,000 species of plants are used by farmers throughout the world, but only 30 species provide 90% of our calorific intake, the top three crops being wheat, rice and maize (corn). Within the preponderant crop species there are many hundreds of thousands of varieties (landraces) adapted to local climates, farming practices and cultural predilections such as taste, colour, structure, storage, etc. Much of this large crop diversity is important for providing starting material for breeding. However, the genetic diversity in crops is much less broad than that of plants or animals living in the wild which indicates the importance of wild species for agricultural breeding programs. There are indications that mixtures of varieties of a crop or of different crops may give higher yields and be more resistant to pests and diseases than monocultures, as shown recently for rice in China. However, even here defined varieties and pure seeds are required.

Human population growth

The world population has risen from 2.5 billions in 1950 to 6 billion today and is expected to reach 9 billion in 2050. 800 million people are currently undernourished. Over 95% of the expected population increase will be in the less developed countries (LDCs) and most will occur in their cities, requiring more living space, water, energy, wood, food and services. For 1995 to 2020 the largest relative population increase (80%) is expected in sub-Saharan Africa rising from 500 to 900 million. Whether the HIV/AIDS epidemic in this region will substantially affect population dynamics is not clear as most of those currently infected are in the reproductive age.

Loss of biodiversity

Loss of biodiversity can be measured by a loss of individual species, groups of species or by decreases in numbers of organisms. The major threats to global biodiversity are, firstly, habitat loss (mostly through expansion of cultivated land and of cities and roads) and, secondly, the introduction of exotic species. Habitats can also be damaged by flooding, lack of water, climate changes etc., natural or man-made.

Since tropical humid forests are particularly rich in biodiversity, their destruction is disproportionately damaging. It is estimated that of the original 16 million km^2 of these forests a century ago only half are left, with about one million km^2 being destroyed every 5 to 10 years. Burning and selective logging may damage an even greater area. Biodiversity is not homogeneously distributed but rather there are areas with particularly plentiful biodiversity ("hotspots") of particular interest for conservation.

Imported plant species threaten the native ones by being highly competitive and often lacking local predators, such as insects or birds. One of the most extreme examples is seen in the pampas of Argentine, a flat grassland with a moderate climate, from which nearly all the native grasses have disappeared and have been replaced by European plants. This invasion was brought about by European farmers, bringing animals and crops, as well as accidentally spreading many different weeds and was already noted in 1833 by Charles Darwin. Islands are particularly threatened by invaders, as is well documented for Hawaii, New Zealand or the Galapagos Islands. Basically, many floras of tropical islands and also floras with lots of relics are particularly threatened.

Biological control agents are often introduced into agricultural ecosystems intentionally to control pests or weeds without resorting to chemical controls. Whilst there have been many welcome successes, such systems may also go wrong. One example is the introduction of the seven-spot ladybird which was intended to control the Russian wheat aphid in the US. This proved to be a competitor of the native ladybird, which then disappeared. The mongoose, an Indian mammal, was introduced to several islands (Fiji, Mauritius, Hawaii) to keep rats and snakes under control: it lead to the extinction of several endemic birds, reptiles and amphibians. Introduced wasps have been seen to kill and possibly extinguish endemic butterflies. In a recent case, a rare and endemic moth has been brought to near extinction with biological control methods. Ever since the silent spring by Rachel Carson warned about the indiscriminate use of pesticides, the conservation community came up with lots of well documented cases of dramatic biodiversity loss in modern agriculture. Many of the "traditional" chemical pesticides are currently under scrutiny and the regulatory system is under revision. It is obvious that reduction of biodiversity in agriculture has also been a reason for invasive weeds to take over.

Are transgenic plants as such prone to spread? In the longest term experiment so far, four different crops (oilseed rape, potato, maize and sugar beet) were grown in 12 different habitats and monitored over a period of 10 years. In no single case were the genetically modified plants found to be more invasive or more persistent than their conventional counterparts. However, one would not expect this to be the case unless the transgene increased its fitness in the wild. There is no plausible hypothesis why crops that have for centuries depended for survival on human care should become weeds because of the addition of one or a small number of well-characterized genes, in addition to the many thousands of genes they already carry due to selection and hybridization over centuries. Large acreages of a single variety without rotation should be avoided with any crop, since monocultures are more prone to disease and pest outbreaks. Organic farming should be reviewed and some of the strategies avoiding chemical pesticides and enhancing biodiversity could well be adopted in modern agriculture.

Conservation strategies

Conservation may be in situ in a more or less natural or habitat or ex situ in some purpose-built environment depending on the particular case. In situ conservation involves the maintenance and protection of natural habitats while botanical gardens and seed banks are used for ex situ conservation.

Conserving a substantial fraction of the tropical rain forests would still allow half or so of their indigenous species to be preserved by selection of the most appropriate "hot spots". Protecting large tracts of land poses major socio-economic and political problems. How forests can be kept free from encroachment by hungry people in search of potential farmland is far from clear. A viable strategy may be to find sustainable livelihoods for rural populations compatible with conserving tropical rain forests. Policing alone will not be successful over vast territories, as seen in the combating of drugs. It should also be possible to evaluate intensivation of agriculture, which is suggested by the United Nations Development Program UNDP.

Conservation also embraces agricultural biodiversity such as crop varieties, land races, semi-domesticated varieties and crop relatives. The role of indigenous communities in maintaining agro-biodiversity is stressed by the Global Biodiversity Assessment and the Leipzig Plan of Action, two recently concluded international agreements.

The Convention on Biological Diversity (CBD)

The United Nations adopted the Convention in 1992 and it entered into force in 1993. This is the first time that a large majority of states, though not the USA, have agreed to a legally binding instrument for biodiversity conservation and the sustainable use of biological resources. A radical change brought about by the CBD is the recognition that states have a sovereign right over biodiversity within their own territory, while previously organisms were considered the common heritage of mankind. Under the terms of the CBD, living organisms or their products may only be removed from a country under mutually agreed conditions. The CBD is a comprehensive approach to biodiversity conservation of both wild and domesticated species at the genetic, species and ecosystem levels. Action is delegated to the national level obliging states to assess biodiversity, enact legislation for its conservation in situ and ex situ, and to enforce legislation within national boundaries.

Biotechnology is particularly touched by Articles 16 and 19 of the CBD since they require a fair and equitable sharing of benefits derived from the use of genetic resources. This includes providing facilities and financial means for technology transfer and open access to scientific and technical information.

The CBD has only been in operation for a few years and so it is too early to assess its long-term effects. As far as biotechnology and "bioprospecting" are concerned, it will take time to establish smooth administrative procedures to allow simple routine implementation of close collaborations. Only if national authorities from countries rich in biodiversity as well as pharmaceutical and other companies see the advantages to be gained from such collaboration will the system expand. The expectations of some countries to make rapid earnings may have been too optimistic. The search for natural, highly active pharmaceuticals in wild plants may often be more cumbersome and costly than modern laboratory techniques using genomics, rational drug design and combinatorial chemistry. One example of a joint effort in "bioprospecting" is the search for specific active ingredients of plants by the Merck Company in the tropical forests of Costa Rica which brought the country 2 million USA dollars over a five year period as well as the potential of royalties if profitable products emerge.

The Cartagena Protocol on Biosafety

The CBD provided a basis for developing an international agreement regulating the trans-boundary movement of living GMOs, called "living modified organisms" in the Protocol, which after a lengthy debate was agreed on in Montreal in early 2000. The main aim of the Protocol is to ensure that GMOs would not endanger biodiversity in the recipient countries: in addition it also covers potential threats to human health. In practice the Protocol will primarily touch the import of transgenic seeds. It requires the exporting country to provide enough scientific information to allow the importing country to judge risks and then to issue, if the product is judged safe, an "Advance Informed Agreement". This would be the basis for an import permit, which is likely to later be taken into consideration when national administrations are asked for a planting permit.

A newly established Biosafety Clearing House collects and distributes relevant information, as defined by the Protocol. Based on the "Precautionary Principle" import licences can be refused, even when there is no clear-cut proof that the product is dangerous to human health or the environment, but such decisions need to be based on a risk assessment. The Protocol is construed to be as important as the World Trade Organization (WTO) agreements which do not allow countries to prevent the import of products unless there are sound scientific reasons showing that the product may cause harm to human health. This apparent lack of hierarchy may lead to some future difficulties. The Protocol will become operational when it is ratified by 50 nations, presumably in 2003. The real effects of the Protocol are impossible to predict at this time: it may promote the use of transgenic crops in LDCs, but it may also make import too cumbersome for seed companies to want to deal with it.

Biotechnology for the acquisition of knowledge

Biotechnology can be used as a tool for acquiring scientific knowledge or to intervene directly in plant and animal breeding and in particular to transfer genetic information from any other organism to a crop or to a farm animal. Today biological research can hardly be conducted without using biotechnology in one way or another. Taxonomy, one of the sciences necessary to study biodiversity, uses molecular markers to identify individual strains of organisms or to identify species in much the same way as is done in forensic medicine to identify criminals. In seed banks genetic fingerprints are used to establish the origin of a seed or the relatedness of plant varieties. A rational classification of most micro-organisms has only become possible with these methods.

Biotechnology is also useful for following genetic markers in plant and animal breeding done by conventional, sexual means. By analyzing a few cells of the newly born calf or of the newly sprouted crop and looking for the presence or absence of certain genes it is possible to predict properties of the progeny which will only show up later in life such as the characteristics of a cow's milk or a crop's expected resistance to a plant disease. Using so-called pre-implantation diagnostics during in vitro fertilization of farm animals, the test can be done even before the embryo is implanted into the cow. These applications of biotechnology are hardly controversial.

Direct gene transfer to crops and farm animals

In 2002 there were about 59 million hectares world-wide planted with commercial transgenic crops mostly in the USA, Canada, Argentina, with smaller amounts in China, Australia, South Africa, Mexico and Spain. Soybean and corn ranked first and second with 37 and 12 million hectares. Cotton and canola accounted for about 7 and 3 million hectares, whilst only small areas of transgenic potato, squash and papaya were grown commercially. Herbicide tolerant varieties dominated with 75%, while insect resistant varieties made up 17%. Virus resistant crops were quite small. USA farmers adopted transgenic crops rapidly because of the economic benefits they offered. In most surveys in different parts of the USA, the yields were the same or somewhat higher with the new seeds. The most noticeable difference to the farmers was the saving on herbicides. Many also found the new crops needed less frequent sprayings and allowed "no till" management. These benefits offset the initial higher costs of the new transgenic seeds.

A large number of transgenic crops are in the development stage, including many tropical crop plants; most of them will only come on the market in a few years, if at all. They are likely to show benefits for the consumers and some may be of particular interest to farmers in tropical countries. Two rice varieties currently being developed should lead to important health benefits. "Golden Rice" contains increased levels of vitamin A and another variety has more iron in the grains than usual, a clear benefit to anemic women and their children. Such rice varieties have been successfully developed in the laboratory but are far from commercialization for both scientific and political reasons.

Several lines of transgenic farm animals have been produced but none have been commercialized. Some lines are made for the pharmaceutical industry to produce drugs in their milk. Others may show improved resistances towards certain infections or produce novel enzymes. Transgenic salmon that grow faster than normal, but reach finally the same size as their wild relatives have been developed and have roused considerable concern amongst ecologists. Here many environmental issues still need clarification.

Native biodiversity and biotechnology

Biodiversity in the wild has been massively reduced in the industrialized countries over a long period and about half the tropical rain forests have already been destroyed. How then can the rest be preserved, given that the world population is increasing rapidly and more food has to be produced regionally?

Yields of cereals in LDCs have gone up very considerably in the last forty years primarily as the result of the Green Revolution. However, the annual growth increases in cereal yields in LDCs have slowed down from about 3% during 1967-82 to about 1% per year from 1993 onwards. These lowered yield increases mean that productivity will probably not keep up with demand. The consequence for biodiversity is devastating as more land will be required for farming which will primarily come from areas with high native biodiversity, in particular tropical rain forests.

The single most promising way to avoid habitat destruction is to increase farm yields in a process that has been called the "Second Green Revolution". Several components will be required including training and education of farmers (in particular women who do most of the farm work in LDCs), more favourable economic and political climates, availability of farm credit schemes, etc. In addition, technical contributions will be necessary and, in particular, improved seed produced either by traditional crop breeding or by modern biotechnology. Reliance will have to be more on the latter since traditional breeding appears to have reached a plateau in yield. So agricultural biotechnology, which is viewed frequently in the public debate as harmful to biodiversity, is, paradoxically, likely to contribute to conserving it by reducing the appropriation of native biodiversity-rich land for farming purposes.

A more limited concern that largely affects Northern Europe is the conservation of native plants and animals, in particular birds, in farmed areas. Their habitats are fields, hedges, roadsides and fallow land where they depend for food on insects and seeds produced by weeds in or near crops. Computer models suggest that more intense weed control measures may lead to smaller amounts of seeds being available to birds. This depends far more on weed management regimes rather than on transgenic plants. Herbicide tolerant beets allow farmers to control weeds later by treating after the seedlings have emerged and long term field trials have shown that those transgenic beets have a positive influence on soil fertility. The more efficient methods may allow setting aside of more land. Farming serves various purposes, particularly in Northern Europe, beyond the primary one of food production. Secondary goals such as conservation of biodiversity and giving pleasure and recreation to city dwellers also need to be considered. Setting aside more farmland requires financial incentives.

Agricultural biodiversity and biotechnology

Could newly introduced transgenic crops transfer genes to native wild plants and thereby change important characteristics of the wild plants? Gene transfer between cultivated and wild plants has always occurred within the limits of species if the two types of plants were in close proximity and flowered at the same time. No new problems can be expected from transgenic plants except if the gene transferred from the GMO to the wild plant significantly increased the biological fitness of the recipient. This generally seems rather unlikely a priori and can, and should, be checked experimentally by risk assessment.

Farmers all over the world use a remarkable number of landraces and varieties of many different crops. These are usually closely adapted to the local climate and topography, and are also used to produce foods for different purposes. In Europe, for example, quite different varieties of potatoes are used for the industrial production of chips from those used for home cooking. Will the traditional varieties and landraces disappear when, and if, transgenic crops are introduced? Will only one or very few varieties dominate in the field? To judge by what has happened on the US soybean market, this will not be the case. Although virtually all herbicide resistant soybeans derive from a single transformation event in Monsanto's RR Soya, hundreds of different varieties have been derived from it by different seed companies using traditional breeding to develop Soya that suit different geographic regions. This shows that at least with this dominant transgenic crop biotechnology has not led to a loss of agricultural biodiversity. On the other hand, judging from the last 100 years in Europe, the diversity of varieties of farm animals and crops has, in fact, gone down considerably, but not because of any biological hazard emanating from some of the breeds, but because farmers need to produce economically. Many old varieties of apples have disappeared, for example, because of the preference of food retailers and consumers. The rapid consolidation in the global seed market may be a concern in this respect and anti-trust legislation may be needed to prevent the dominance of few oligopolies.

Could landraces be threatened by gene transfer from GMOs? Gene transfer by pollen has always occurred between different landraces and between different new varieties of crops. Despite this, varieties of cereals or fruits have been stable over many years and specific traits have not disappeared. However it is important to store landraces and native relatives of crops for their intrinsic value as well as for having starter material for future crop breeding. The method of analysis using gene probes, which is no longer solely dependent on visible traits, has become far more precise.

Gene transfer from crops to their wild relatives has to be evaluated case by case and region by region. Field experiments have shown that when hybridization rate is high, the probability of transgene outcrossing is also high. This has been shown in a Swiss study with a Dutch – Swiss evaluation system, which accounts for three major factors in gene flow, namely hybridization rate, frequency of crops and their wild relatives and also dispersal characteristics of the seeds.

Gene transfer from crop to crop can also cause problems: There are farmers, especially in the developed world, who do not want to have transgenes intermixed in their fields. Seed producers have learnt over decades to produce seeds with a well defined genome.

Problems might arise from the fact that the new screening methods for standardized seed genomes are much more sophisticated, since it is now possible to screen transgenes with high exactidude. Tolerance levels such as the adopted 1% in some countries would be the solution. A way out of the unwelcome crop to crop transgene intermixture would be to produce biological barriers for the transgenes to hybridize, such as apomixis (spontaneous embryo development without crossing through pollen transport) or many other means.

Economic considerations

Under what economic conditions will biotechnology benefit agricultural productivity and thereby help biodiversity? The original green revolution which improved wheat and rice production in Asia was initiated by research carried out in the public sector. Today's innovative research in biotechnology is mainly done by a few large companies with modest contributions from public sector and small companies. Companies strive for intellectual property rights (IPR) through patenting for return on their research investment. This is largely unavoidable with the present IPR systems. Provisions need to be made so that the agricultural research institutions of LDCs obtain access to information as well as patented materials and procedures needed for regional farmers.

A limited number of transfers of specific patents has already occurred from companies such as Monsanto and Syngenta for producing virus-resistant sweet potatoes and Vitamin A enriched golden rice. This approach, including the possibility of enforced licensing, is probably more practicable than changing the world's patenting system which does, however, need to be adapted in order to deal with living organisms. A new issue arising in the field of IPR is the attempt to protect traditional, indigenous knowledge through novel international legislation.

Lack of food appears to be, and is largely, a problem of distribution; there is not sufficient food available to those who need it. The poor simply do not have the money to buy enough. Since in LDCs most of the poor still live on the land, it is here that production has to be increased so that there is not only enough for survival but a surplus which can be sold. Increasing local agricultural productivity is the most effective way to do this and should, according to many economists of the public sector including those of the UNDP, embrace modern biotechnology as an important component.

Conclusions

The issues involved in the interaction between biodiversity and biotechnology have far-reaching consequences and need to be subject to an open and knowledge-based dialogue in society. The dialogue needs to include many different stakeholders, including farmers of LDCs, diverse scientists, policy makers and communicators. Cultural values involved in farming and food production need to be taken into consideration, just as much as the emotive aspects of eating and drinking.

Agricultural biotechnology may have quite different impacts on biodiversity, depending on what particular application is considered and how it is applied. The result will depend both on the agronomic as well as on the societal context. If in the LDCs, new transgenic seeds are only available to a small group of large farmers, this may lead to large monocultures and a modest improvement in sustainability with little alleviation of poverty. However, the technology in itself is well suited to small farmers, since it is packaged in a small seed grain. One of the best examples can be demonstrated in China: Hundreds of thousands of small farmers have adopted Bt cotton with good results. The application does not depend on large amounts of capital to buy expensive machines like combine harvesters. However, if the seeds are to reach small farmers, there need to be many seed outlets (public cooperatives or private companies) and adequate agricultural extension programs.

The most surprising outcome from this review is that biotechnology has the clear potential to save native untouched land, particularly in tropical rain forests, by reducing or eliminating the pressure to bring more land into cultivation. Increasing productivity on land that is already farmed is the best hope to save some of the land with the richest native biodiversity, the tropical rain forests. This central idea, leading to improved sustainability and reduction in poverty, needs to spread further and to be embedded in ethical considerations. The UK Nuffield Council on Bioethics concluded that because of this potential of agricultural biotechnology "The moral imperative for making GM crops readily and economically available … is compelling." Similar conclusions were reached in a joint report put out by the Science Academies of India, China, Mexico, the United Kingdom, Brazil and the USA as well as the UNDP stressing that there is a clear need for this technology. These and other sources from the public sector maintain that agricultural biotechnology has the potential, when judiciously applied to increase farm productivity amongst the poor in the LDCs. As a consequence some native biodiversity can be expected to be saved. See also the website of the Royal Society of New Zealand, probably the most comprehensive collection of arguments for and against transgenic plants and derived foods.

References

Transgenic Plants and World Agriculture, Royal Society, July 2000. http://www.royalsoc.ac.uk/ policy/index.html Report der Royal Commission of New Zealand, http://www.gmcommission.govt.nz/

Agricultural Biotechnology and the Poor, G. J. Persley and M. M. Lantin (eds.), US National Academy of Sciences Consultative Group on International Agricultural Research, Proceedings of an International Conference, Washington D.C, 21-22 October 1999, http://www.cgiar.org/biotech/rep0100/contents.htm

Reviving the stalled momentum of global poverty reduction: what role for genetically modified plants. M. Lipton, 1999 Sir John Crawford Memorial Lecture, CGIAR International Centres Week, CGIAR Secretariat, http://www.cgiar.org

Genetically modified crops: the ethical and social issues, Nuffield Council on Bioethics, May 1999), http://www.nuffieldfoundation.org/bioethics/publication/pub0010805.html

Human Development Report 2001, United Nations Development Programme: Sustainable Development. Press Kit: July 10, 2001, http://www.undp.org/hdr2001

ISAAA: 2002 Global GM Crop Area Continues to Grow, http://www.isaaa.org

Session 1: Impact of agricultural biotechnology on ecosystems

Methods for Risk Assessment of Transgenic Plants
IV. Biodiversity and Biotechnology
K. Ammann, Y. Jacot and R. Braun (eds)
© 2003 Birkhäuser Verlag Basel/Switzerland

Widening perspectives on biodiversity

Anatole F. Krattiger

SWIFTT, International Programs, Cornell University, 31A Warren Hall, Ithaca NY 14853, USA

Introduction: Point of departure for a wider perspective

It is perhaps symbolic, and certainly appropriate, that a conference of this importance should begin with a talk that aims at "widening our perspectives on biodiversity". An attempt to widen perspectives on biodiversity is, in my opinion, long past due. All of us at this conference recognize the importance of biodiversity. We all want to enjoy and to conserve our resources, yet biodiversity has somehow become an issue that "environmentalists" claim for themselves. Our goal, however, should not be to take the issue away from environmentalists, but to ensure that more voices from more constituencies are included in the discussion about biodiversity. This is a precondition for a more effective approach to the conservation, sustainable use, and equitable sharing of the potential benefits of biodiversity. But where do we begin? The topic of this conference is forbiddingly broad. And consider the subtitle: "Proposals for Action." This further complicates matters. To act successfully we must think past the individual subject matters within the complex sciences of biotechnology and consider the other multiple sciences and social and economic factors that affect biodiversity. We must also, I dare suggest, step outside the province of science itself. Biodiversity is not a purely technical and scientific issue. For concrete and effective proposals for action, we must think in an all-encompassing way that invokes both traditional concepts of human culture and development, and those of environmental stewardship, which in my view are not two opposing principles, but complementary ideals that should guide the creation of our vision of the future and the steps we take to reach it.

Ralph Waldo Emerson once wrote: "This time, like all times, is a very good one, if we but know what to do." Do we know what to do today to preserve biodiversity and promote human development? What proposals for action are we able to consider here? What proposals for action do we dare consider? And is our vision wide enough? Because where our vision is limited, our actions will be circumscribed. Allow me to outline one personal view.

A few weeks ago I was scuba diving in the Caribbean, exploring the underwater world's rich biodiversity, viewing spectacularly shaped corals inhabited by vibrantly colored fish and an incredible diversity of species unknown to me. It was, to some extent, a mysterious and even mystical experience. Deep down in the ocean, I felt that in the colors and shapes and movement and life that surrounded me I had come close to the source of all true art – and science. I paused in wonder, floating rapt in awe at the shimmering world around me.

At one time or another we have all asked ourselves how the complexity and beauty of the world's biological diversity could have emerged and evolved from life in the sea. Today, thanks to the sciences, we all know the answer: through the process of evolution life became richer and more complex as millennia went by. Nature crafted the world we live in, giving us all the raw materials – and through the application of modern biotechnology the "templates" – for more prosperous, safer, and healthier lives.

The biotechnology potential for environmental conservation

Biotechnology, in the "environmentalist's" quarters, is often regarded as a threat to biodiversity and plenty has been written and said about that subject: some of this writing is speculative, some scientifically accurate and important and some merely is a way of drawing attention to someone's cause without any relevant truth. We have seen little comparative analysis, however, of the potential of biotechnology to benefit nature and biodiversity.

Yet is nature herself who first invented and applied biotechnology. Nature used it dynamically, enabling new species and forms of life to emerge and to evolve. Some species took over, others disappeared. But life for each and every species was not sustainable – only progress and evolution was sustainable.

Even humankind is a product of that evolution and in my opinion it was high time that we discovered – not invented but discovered – biotechnology. So much good can be done with it, if we apply and deploy it safely, wisely and cautiously. Caution must not be our only guiding principle, however, for we all know the pressing human needs that face us. Hunger, malnutrition, and poverty stalk most of the world's population, and dire environmental consequences threaten if we fail to protect nature and change our methods of production, especially our agricultural production. We need to get biotechnology right.

How then can biotechnology make the world we live in better? First, let me say that biotechnology is already making a difference. But consider what biotechnology can do today for people from all walks of life and in many countries of the world. Thanks to biotechnology, improved crops that require less pesticide, herbicide, and that have better post-harvesting characteristics are already being grown on 50 million hectares, making it the fastest agricultural technology to ever be adopted.

In summary, biodiversity benefits from biotechnology directly, for example through better science for conservation, but more importantly, biotechnology indirectly benefits biotechnology. This becomes evident if we look at seven major constituencies, their objectives and environmental impacts (see Table 1).

At the end of the day we all want the same things. We could – and must – do much more to reach these goals. The question is only: Where do we start?

The same thinking can be applied to biodiversity conservation. We all want it. Indeed, I hope that this conference will not only produce good, new ideas about how to preserve biodiversity, but that it will also prioritize them. We don't have time or resources to waste. In my view, biotechnology offers important options to conserve biodiversity, especially since it can help limit destructive agricultural practices and make it possible for farmers to produce more food on less land.

Now, let me attempt to ask a few more questions, some challenging, I hope, and some controversial, I'm sure, about six major areas where biodiversity and biotechnology intersect.

The bottom line is that with biotechnology, we allow agriculture to become more productive and use more environmentally sustainable technologies with less impact on the environment.

The nexuses of biodiversity and biotechnology

I want to look at five central areas and make some bold propositions on the nexus of biotechnology and biodiversity. All proposals and points are aimed at challenging you and providing some points of departure, albeit controversial ones, as a way of getting the thinking and discussion started.

Table 1. Objectives and environmental impacts of seven major constituencies

Constituency	Objectives/goals	Environmental/development impact
Farmer	High yielding crops	Productivity increase
	Lower use of pesticides	Decreased environmental impact
	More efficient use of inputs (e.g. water, less transport)	Decreased secondary environmental impact
	Produce high quality products	Productivity increases, value added, healthier crops
Plant breeder	Better tools to make their work more efficient	Scientists in both developed and developing countries are now able to breed disease-resistant, delayed-ripening, and hardier varieties of crops.
	Produce varieties that farmers want and need	
	Provide consumers with more nutritious and all-around better varieties	
Scientists	To advance the frontiers of science	Progress has been made, especially in medicine, where disease diagnostics and the production of substances that were too expensive, such as insulin, are now allowing people everywhere to lead healthier lives.
	Discover and invent exciting new technologies	
	Benefit humankind	
Consumers	Plenty of food with good and increasingly better nutritional value at relatively affordable prices.	Biotechnology is making progress in that direction possible – these are the most exciting applications in the research and development pipeline.
A CEO of a company, large or small	To sell what the consumer needs in a way that ensures that the company is sustainable and profitable	The economic potential of bio-technology is enormous, for both industrialized and developing countries. It can be a win/win proposition if we work together and resist fear and distrust.
Government officials	To ensure that the people of their country equitably enjoy the benefits of science and technology	Biotechnology could help worldwide, if only the minister were a true friend of the earth!
	To cast out the image of poverty haunting their country	
Mothers and fathers	To ensure that their children can enjoy a better life on this planet	Biotechnology is helping already, in agriculture, health, and the environment, to make the world a better place.

Agriculture and population

"Certainly no other area of crucial importance to the future of civilization has elicited as much mindless comment as have human demographic patterns". Paul Ehrlich once said that.

I do not in any way wish to add to the list of such mindless comments. Instead, let me point out that today the power of politics has trumped the truth of science. Here, certainly, is one area that requires proposals for action: We must somehow better communicate the science of biotechnology to the public and to politicians. When it comes to biotechnology, the world is drowning in "information" but thirsty for knowledge.

But how do we cope with the environmental luddites and anti-industry activists, who claim to be clearing things up as they misquote, misrepresent, misunderstand, or knowingly fail to read the scientific literature? How should we respond to their continuing propagation of error-riddled rhetoric? Perhaps we should we go back to the Brundtland report of the mid-1980's on Sustainable Development and only use the politically correct stance of anti-over-consumption and pro-sustaining sustainability.

Even those who see biotechnology as somehow a potential "problem" at least acknowledge that today there are real population/resource problems. Yet many still fail to recognize that agricultural biotechnology is an imperative for the 2.4 billion people – the 40% of us – who survive on less than TWO dollars a day. They somehow fail to see the environmental benefits of agri-biotechnology, which can increase the productivity of existing farmland and thereby reduce the impact of agriculture on the environment, particularly on marginal and fragile ecosystems.

The critics of biotechnology are doing little to meet the developing world's growing food needs, and, in a paradoxical twist, this will result in even more environmental destruction. We need to make clear to people – such as Prince Charles – that the latest environmental catastrophe in Mozambique will seem like a Royal Garden party compared to the consequences of the social shifts and consequent environmental threats to come. Denying the poor access to the benefits of biotechnology will not provide them with the means of lifting themselves out of poverty. Denying technological advances is surely one way of sustaining subsistence farming which farmers don't want and which doesn't help the environment either.

Globalization and trade

Back to the Brundtland Report. It marked the turning point from industrialization to post-modernity. With it, today's "international" debate and powers of trade are characterized by their recognition that there are limits to growth, that intergenerational equity ought to have a place in modern socio-politics, and that there is a global interdependence, stretching from the environment to societies to economic affairs. Now what? How has biodiversity benefited from such high words?

Granted, this debate has established valuable principles but what action should follow from these principles? Given that resources are not evenly distributed across the globe, that productivity in certain crops is much greater in some areas than in others, and that needs are different from country to country, it is clear that increased liberal trade should be the primary element in a strategy that aims to make the best use of scarce resources because trade will lead to the most efficient use of resources wherever they are. Yet we leave trade – and trade barriers – to the politicians at the World Trade Organization. It is too important a source of wealth for that.

And let us look at food labeling as a trade barrier. Who will make the decisions for the world? Wealthy, middle-aged, white, well-fed, well-clothed, well-educated activists in Seattle? Let's remember that 80% of the world's consumers don't care about labeling – they simply want food on the table. And of these, nearly 40% cannot read anyway. In any case, the majority of foods sold to local consumers comes without packaging. Who is representing the voices of the hungry and poor during such debates?

This gives us one more criteria about our proposals for action: We need proposals that begin with the world as it is rather than proposals based on a romantic image of how the world should be. Feeding the hungry and preserving biodiversity won't be accomplished merely with organic farming. We need to stop holding more than half the world's population hostage to poverty under the guise of debates about "safe" foods or "safe trade." These "debates" are more about lifestyle choices in the industrialized, Western countries than they are about science-based analyses of new agricultural technologies or of socio-economic realities affecting the 2.4 billion poorest of our world.

Preserving biodiversity is not an elitist enterprise. After all, aren't the first victims of environmental degradation the poor? In fact, they are the ones already suffering, yet they have had and continue to have no voice in the debate about either agricultural biotechnology or the establishment of new international regimes such as the biosafety protocol.

Bioprospecting and conservation

Few would hang conservation on so materialist a thread as cash returns from bioprospecting rights. But the ethical, esthetic, religious, community-sustaining, and other values called upon to support conservation are not moving finance ministers and development experts. Their initial response is to see progress only in terms of the bottom line. Of course, all species and all human cultures that have extracted meaning from this life are worth protecting, if for no other reason, some would argue, than for the postmodern metaphors they still have not revealed. Not surprisingly, therefore, we may be better off to spend resources on conceptualizing, describing and defining the "intrinsic" values of biodiversity, and then "selling" the ecological, ethical, and esthetical values of biodiversity. Instead, we have focused on devising systems of access, on control, on sovereignty, brought about largely by the Convention on Biological Diversity. And this division of ownership, in my view, has become part of the problem rather than the solution.

But what institutional arrangements can ensure that benefits are "equitably" shared between companies and countries and individuals and people? How can we even define "equitable sharing" when equitable is so much a concept that depends more on the eyes of the beholder than on measurable characteristics? Clearly, we need to bring about shifts in the perception of values, including prejudices about modern biotechnology. Perhaps a new type of biotechnology enterprise, one publicly owned but that is managed like a private enterprise, would allow us to sway public opinion. We certainly need something big – and bold – if we are to bring about needed change. Let Monsanto buy the CGIAR. Or should it be the other way round?

Intellectual property rights and technological innovation

One of the fundamental problems is that the knowledge revolution that biotechnology and biodiversity have enabled has not been followed by a revolution in intellectual property (IP). Once a sacred right of the inventor, IP was stimulated both by metaphysical arguments over ownership and by a desire to take practical measures to make inventions quickly available. But times have changed. Consider Linux, the open-source software to which everyone can contribute. It has seen the fastest software evolution ever.

At first Linux was developed as a response to Microsoft Corporation's domination of the operating system world. It was driven by this ideology, but it is making business sense too. The collaborative efforts of programmers from around the world have created an impressive software operating system that is gaining more and more market share. This "open-source," group effort has turned business assumptions on their head: people now see that the value in software is not the software per se, but the productivity gains it allows.

The same might apply to DNA and genes. One might well argue that we need an open platform where everyone can contribute, where everyone benefits. What a difference this would make if everyone could contribute to and benefit from the productivity gains enabled through a better understanding and knowledge of DNA and genes!

What proposals for action make sense? What would be needed to make such an "open platform" happen is a new definition of patents to allow for continued investment in the science and technology and products. Perhaps a working group to try and come up with something new is what is needed, and would also save the companies hundreds of millions of dollars in legal fees. Yet today we are bogged down in discussions about the morality of ownership. What we need is less dogma and more common sense.

International organizations and regimes, and the precautionary principle

Here I will only say that I believe the relentless push for new international treaties and protocols is counter-productive. What we really need to find are better economic instruments. But let's leave that subject, and come back to the precautionary principle, used and abused by many NGOs as a weapon of last resort against progress and as a way of pushing for new international treaties, like the Carthagena Protocol (or the Biosafety Protocol under the Convention on Biological Diversity).

The precautionary principle… have we truly understood its meaning? It was formulated and defined in the 1992 Rio Declaration on Environment and Development as: "… lack of full scientific certainty shall not be used as a reason for postponing cost effective measures to prevent environmental degradation". In other words, where there are threats of serious or irreversible environmental damage, lack of full scientific certainty shall not be used as a reason for postponing measures to prevent environmental damage.

Taking this definition to its logical conclusion must mean: "Where there are unquantified and unqualified potential risks, such as those of biotechnology, the lack of full scientific certainty shall not be used as a reason for postponing measures that can prevent environmental damage with the deployment of biotechnology." Isn't it becoming increasingly evident that biotechnology as applied to agriculture leads to fewer pesticide applications in many cases and hence more and better foods. Consequently, should the precautionary principle be applied in crops that demand more pesticides?

What other proposals for action are appropriate here, if not proposals that demonstrate the positive impact of biotechnology? Maybe then we can influence ethical views and act on recognized, basic human values. One such basic human value, I trust we agree, is health.

Even Europeans accept biotech for pharmaceuticals. But in developing countries the biggest health hazard is malnutrition. It kills 40,000 children per day! So what is a luxury for Europeans (improved, safer, more plentiful, and cheaper food) is a matter of survival for the majority of the inhabitants of this earth. Enough food means access to basic health. And we are denying it to most of the people on this planet!

We think we are opposed to new solutions for ethical reasons. Yet if one analyzes the reasoning of most people, it boils down to irrational fears, inherent opposition to change, and anti-science beliefs. I dare say that we, the scientific community, are probably only dimly aware of the fundamental "anti-science" systems of belief so widespread today, from harmless new age crystal power to politically dangerous creationism, without realizing that perhaps, and only just perhaps, both creation and evolution are possible. In either case, what we need to demonstrate is somewhat more respect for the opinions of others and tolerance.

Paradoxically, however, there is a widely accepted modern world-view that includes science as a central element. Indeed, everyone demands scientific certainty in respect to biotechnology's impact on biodiversity. How can that paradox continue to exist? How can nearly everyone hold one or more views incompatible with the scientific ideal that is the "gold standard" for obtaining answers to our questions about the nature? Evidently, we all seem to have a high degree of tolerance for contradictions.

Such contradictions also appear elsewhere in debates about biotechnology. Our market system has failed because we continue to build perverse economic incentives rather than remove disincentives. And then there are the failures and contradictions of government intervention, in which agricultural subsidies perversely damage the environment and place the poor at the biggest disadvantage.

Yet despite the fact that we have known about these contradictions for years, we haven't been able to come up with pragmatic proposals for action. Indeed, one of the big dangers ahead is that we increasingly depend on international relations, international institutions, and international politics as the means to solve environmental problems. Politics has become a management tool. At the same time, global environmental management is pure politics, a practice that too often results in forests going up in flames.

What happened to the precautionary principle? Why do we continue to proceed with the status quo if it leads to such destruction? If we wait for absolute scientific certainty, it will be too late to design effective legal, political, and economic responses to the pressing problems of today. Yet how should we proceed?

Conclusions

I challenge everyone to think of new proposals for action, keeping in mind a broader, wiser, more ingenious, and especially more daring and global approach. Sometimes we need to go back to the origin of things, to the root cause of our problems, maybe even to the origin of life.

I began my talk by describing the awe I felt at the mysterious life in the sea off Barbados. Let us keep our eyes open to the awe of the possibilities that biotechnology has put in front of us. Let us seize the opportunity to make a difference. And remember what Albert Einstein once said (who had worked here in Bern as a patent agent): "The most beautiful thing we can experience is the mysterious. It is the source of all true art and all science. He (or she) to whom this emotion is a stranger, who can no longer pause to wonder and stand rapt in awe, is as good as dead: his eyes are closed."

Let us keep our eyes and mind wide open for the next two days during our conference.

Methods for Risk Assessment of Transgenic Plants
IV. Biodiversity and Biotechnology
K. Ammann, Y. Jacot and R. Braun (eds)
© 2003 Birkhäuser Verlag Basel/Switzerland

Horizontal gene flow

Sir John E. Beringer

Senate House, University of Bristol, Tyndall Avenue, Bristol, BS8 1TH, UK

Introduction

There is a concern that genes may flow from genetically modified organisms (GMOs) to sexually compatible relatives (vertical gene flow) and to totally unrelated species and families of organisms (horizontal gene flow). This concern is based on an entirely erroneous feeling that cloned genes are intrinsically more likely to be involved in horizontal gene flow that other genes in chromosomes.

Gene transfer between crops and sexually compatible relatives has occurred since the domestication of plants first began. Initially such gene transfers were among plants that shared the same gene pools and thus were unlikely to produce progeny that would seriously threaten the gene pools of the wild species. Over the centuries farmers kept seed from the best plants in their crops that had been formed by mutation, or had arisen from natural crosses. Gradually major differences arose between the domesticated and wild species, so that farmers were keeping alive plants that contained combinations of genes that improved the domestic, but not necessarily the environmental, attributes of the crops. In more recent years plant breeders have broadened the range of plants used for breeding and have introduced genes from relatives that could not naturally have cross-pollinated the crops concerned. In this way the rate at which novel genes are entering crop species, and hence are moving from crops to wild relatives, has been greatly enhanced.

The transfer of genes from genetically modified (GM) crops will continue the process of adulterating the gene pools of sexually compatible relatives. When the genes that are transferred from GM crops are derived from different species, or indeed families, gene flow from the crops can be seen as a method whereby vertical gene transfer leads effectively to horizontal transfer of the cloned genes to the wild relatives. In this way the flow of cloned genes will be enhanced, but will be contained within the interfertile species undergoing normal vertical gene flow.

Genes flow from crops can alter wild relatives in areas in which crop production has been widespread. Examples of where gene flow has affected relatives, and hence biodiversity, are hard to find (see Gray, this volume). The work of Al Mazyad and Ammann (Al Mazyad & Ammann, 1999) showing that vertical gene flow from *Medicago sativa* to *M. falcata* is leading to the rapid decline of the wild relative in Switzerland provides an important warning of what might be happening elsewhere. However, it also raises important questions. If we know that a crop is affecting the biodiversity of cross-pollinating relatives, do we restrict growth of the crop? If we are to restrict the growth of crops on the basis of gene transfer risks, how do we decide which relatives are sufficiently important to apply this degree of intervention of agricultural practice? The harm being caused to populations of *M. falcata* is by gene transfer from a traditionally-bred crop, but what would be the public response if by chance the first well documented report of this nature had been about gene flow from a GM crop?

The transfer of genes between domesticated animals and related species has been much less frequent due to constraints on interspecific mating and the fact that for most livestock close relatives are not usually present. Fish are an exception, but at present intensive breeding, other than for ornamental fish, has been limited. On the other hand, horizontal gene transfer among microorganisms is probably relatively common and natural procedures are known whereby gene flow can perhaps occur among all species of bacteria.

Methods of gene transfer

Gene transfer within species is almost essential to preserve the fitness of most species of animals and many species of crop plants. It is also very common among microorganisms. Most school children are fully aware of the methods used by animals and plants and little needs to be said here about these processes.

For all methods of gene transfer there is an absolute need for the incoming DNA to be replicated if the genes carried on the DNA are to be maintained in the new host. This can result from the integration of the DNA within the genome of the host, or by the DNA being replicated independently (such as plasmids in bacteria). Integration usually requires that there is a very similar sequence of DNA in the host so that the incoming DNA can "pair with the region" and undergo recombination. Occasionally DNA molecules contain sequences of DNA (called insertion sequences) that are capable of bringing about "non-homologous" recombination. Replication of nucleic acid molecules without integration into existing DNA molecules depends upon the presence of sequences that are recognised by the host as sites to start and end replication of the nucleic acid concerned so that it can be maintained as an independent genomic element.

If DNA that has entered a cell is to be maintained there is a need for it to confer a selective advantage on the host, or be very closely linked to a gene of this nature. If it does not, the frequency at which genes are present in populations will remain at the frequency at which the genes are inherited. Thus, although some species are very effective in taking-up DNA from the environment, they remain species because integration of foreign DNA is very infrequent and seldom does such integrated DNA confer a selective advantage on the new host.

Other factors that are likely to reduce the frequency with which DNA can be maintained in populations are that different species can have different regulatory signals controlling the expression of genes, and also gene expression can be affected by different preferences for codon usage.

Genes can be transferred among species of bacteria by three different processes: transformation, transduction and conjugation.

Transformation is the uptake of naked DNA and integration of that DNA within the genomes of host bacteria. Transformable species have special mechanisms to facilitate DNA uptake in a form in which it can be recombined into the genome. Usually competence to take up DNA is transient and dependent on the stage of growth. Transformation potentially offers the widest possible opportunity for the transfer of genes both among microorganisms and from plants and animals to microorganisms. As stated above, the uptake of alien DNA is certainly possible and surely occurs, but is seldom likely to result in the inheritance and expression of alien genes. Indeed, genes from eukaryotic organisms are likely to contain introns whose presence will prevent expression of the genes in bacteria.

Transduction is the movement of DNA among bacteria carried within the heads of bacterial viruses (bacteriophage). Two forms of transduction are known. Restricted transduction, in which viruses that are integrated within bacterial chromosomes are replicated and packaged with occasional errors such that genes adjacent to the bacteriophage within the host genome are transferred. Generalised transduction is when any region of DNA within the host genome is transferred through mistakes in packaging the heads of the bacteriophage. Both events are relatively infrequent, as they are errors in natural events during the production of bacteriophage. Furthermore, neither is likely to lead to significant horizontal transfer of genes because the host range of bacteriophage tends to be limited.

Conjugation, the pairing of cells and transfer of genes, is the most likely method by which large regions of DNA can be moved among bacteria. Conjugation is brought about by plasmids, which are usually circular DNA molecules that replicate independently from the chromosome. They code for genes that initiate pairing, which is determined by the host bacterium, so that gene transfer for some types of plasmid does not need recipient cells to have specific genes that enable them to act as sexual partners, or as recipients of plasmid DNA. Indeed, there are plasmids which code for transfer genes that enable gene transfer to occur between almost any bacteria and even from bacteria to plant cells (Buchanan-Wollaston et al., 1987). Conjugation is primarily an activity that enables plasmids to move among potential hosts. In order to be maintained in new hosts, plasmids carry genes that can confer a selective advantage on the hosts concerned. Under strong selective pressure, such as during the administration of antibiotics, selection for bacteria that have inherited antibiotic resistance plasmids is very effective, compromising therapy.

As well as moving plasmid genes, conjugation allows the transfer of genes from chromosomes. Transfer can occur through the interaction of plasmids with host chromosomes during which fragments of chromosomal DNA are transferred. These fragments can carry many hundreds of genes into a new host, unlike the situation with transformation and transduction in which a hundred gene fragments would be large. The interaction with host chromosomes can also result in the formation of plasmids in which sequences of host chromosome become integrated within the plasmid (plasmid-primes). The inheritance of fragments of host DNA mobilised by plasmids will depend upon the DNA being homologous with DNA in the new host to enable recombination, or the presence of insertion sequences to facilitate non-homologous recombination, or the presence of origins of replication to enable independent replication, as discussed above. However, if the fragments of DNA are integrated within plasmids (plasmid-primes) they will be replicated within the new host as part of the plasmids, providing an excellent way in which wide intergeneric transfers of DNA can become stabilised. Whether or not genes on plasmid-primes are maintained will be determined by the functions coded by the genes concerned, or those of the core plasmid itself.

An important characteristic of GMOs that needs to be more properly recognised is that there is no reason why transgenes should be more likely than any other genes to spread horizontally. Indeed, as it is possible to ensure that they are not associated with insertion sequences, or DNA sequences that allow their independent replication, they should be less likely to spread than genes inherited by traditional means from wild relatives. Thus, when assessing possible risks to biodiversity from horizontal gene transfer much the greatest threat is not transfer of the genes to wild relatives, but from the nature of the released GMOs. For example, the very successful development of salt tolerant crops could lead to ploughing and other forms of weed control in areas presently largely undisturbed by agriculture. Conversely, if GMOs enhance productivity on existing agricultural land they might reduce the spread of agriculture and related adverse effects on biodiversity.

Horizontal gene transfer

Plants

Interspecific cross-pollination is quite common, but seldom results in progeny that are sufficiently fit to compete with and displace either parent (Raybould & Gray, 1993). If this was not the case speciation as we know it would not be observed. True horizontal gene transfers have occurred during the evolution of life (Lin et al., 1999; Mayer et al., 1999), but a great deal more genome sequence information will be needed to separate distant horizontal gene transfer from very high degrees on conservation of sequence. Looking from an evolutionary time scale, the most important horizontal gene transfers have been the inheritance of plastids and mitochondria and the resulting migration of some of the genes from the organelles to the host chromosomes (Andersson and Kurland, 1998; Lin et al., 1999; Mayer et al., 1999).

Gene transfer from bacteria to plants is a well-known natural phenomenon and forms the basis for much plant genetic manipulation. The bacteria concerned (*Agrobacterium* species) have evolved a series of plasmid-borne genes that enable them to attach to exposed cells in wounded plants, transferring genes from the plasmid to apparently random sites within the plant genome (Fraley et al., 1986). Once integrated, the genes cause the plant cells to produce nutrients that agrobacteria can preferentially utilise. This is a particularly unusual form of horizontal gene transfer in that it is frequent and is an essential component of the pathogenic relationship between the host and microorganism. Infection by *Agrobacterium* is not a means whereby the genes become stably inherited, because infection is through somatic cells and the bacterial genes are not seed transmitted. We have no evidence to date that other bacteria have evolved specific methods to transfer genes to plants, or animals.

There is a great deal of interest in the possibility that genes will transfer from plants to bacteria; particularly to species that have very effective uptake mechanisms for foreign DNA. Attempts have been made to demonstrate transfer of cloned genes that can be expressed in bacteria, and for which strong selection can be made (e.g. kanamycin resistance). However, even using *Acinetobacter calcoaceticus*, which has evolved a DNA uptake process that does not discriminate plant and bacterial DNA, evidence for horizontal gene transfer has yet to be obtained (Neilsen et al., 1997). Unconfirmed reports that similar work looking for the inheritance of an ampicillin resistance from transgenic maize has proved negative reinforce the expected: horizontal gene transfer from plants to bacteria is possible, but is likely to be very infrequent.

In fungi there is less evidence for horizontal gene transfer, although there has been some concern about the ability of *Plasmodiophora brassicae* to take up plant DNA. This fungus has evolved processes that enable it to take-up large amounts of DNA from degrading brassica plants. Buhariwalla and Mithen (1995) have looked for evidence for the inheritance of plant genes, but have found no evidence for the presence of transgenes in resting spores. Occasionally small amounts of repeated brassica DNA sequences are observed. As these are related to transposable elements, it might be expected that they could become transiently integrated, but not inherited, or they would be found in all isolates of the fungus.

Another form of horizontal gene transfer is that of viral nucleic acids to host plants, such as the multiple repeats of gemini virus DNA found in tobacco genomes (Bejarano et al., 1996). As interactions between viruses and host nucleic acids are well known for many groups of plant, animal, and microbe viruses, it is perhaps self-evident that transgenes carried on some groups of viruses are potentially highly likely to be integrated within the genomes of host species. We need only look back to the early days of microbial genetics research when studies of the genes coding for lactose utilisation were facilitated by the natural ability of the temperate bacteriophage to carry *lac* genes among strains of *Escherichia coli*.

Microorganisms

As discussed above, there is excellent evidence for the ability of bacteria to undergo interspecific gene transfer, but what evidence is there for such transfers? It comes from genome sequence information (Andersson and Kurland, 1998) and from various genetic studies (Fry and

Day, 1990). As more and more bacterial genome sequences become available it is clear that there are regions of different DNA base composition in most (all?) bacteria indicative of past distant crosses. Clearly, one would expect plasmids to move widely, but it is perhaps less expected that chromosomes would contain sequences from different bacteria so unrelated that the DNA base sequences are obviously different. However, I believe that this should be expected because bacteria are very small and very large populations are quite common, enabling even rare events to occur. Plasmids can be transferred naturally at frequencies greater that 1%, making it likely that gene transfer is the rule, rather than the exception for some species of bacteria that routinely form large mixed populations, such as those growing in and on animals or plants (Lilley et al., 1994).

A particularly good example of the extent of gene transfer among bacterial plasmids is provided in work by Touch et al. (2000). They report on a plasmid carrying resistance genes to 16 antimicrobials that was isolated from a human pathogen, *Corynebacterium striatum*. It contains resistance genes first observed in the following bacteria: *Corynebacterium glutamicum, Erwinia amylovora, Pastuerella piscicidia, Corynebacterium diptheriae* and *Mycobacterium tuberculosis*. This example shows the potential for plasmids to act as foci for the bringing together of genes from different species, presumably under conditions of strong selection during antibiotic therapy. Homologous recombination appears not to have been required for the different genes to come together because the plasmid contains eight insertion elements that could have been involved in bringing about illegitimate recombination. Somewhat unexpectedly the genes come from bacteria which inhabit very different environments; soil, plants and animals, so we should assume that the potential for bacteria from different environments to inherit each other's genes is realised in Nature.

An earlier study of the development of penicillin resistance in streptococci showed the formation of hybrid genes derived from interspecies transformation of fragments of chromosomal DNA from relatives having close DNA homology (Maynard Smith et al., 1991). For these bacteria, penicillins are widely used to control the diseases they cause (gonorrhoea and respiratory infections) leading to almost constant antibiotic selection for the development of resistant strains. A practical demonstration of the significance of the emergence of drug-resistant gonococci is that it affects the ports at which it is wise for naval ships to dock!

Antibiotic resistance provides a good example of a characteristic for which selection is greatly enhanced by recent advances in medicine. It demonstrates that horizontal gene transfer can provide a rapid and effective mechanism to share useful genes within populations of organisms that can exchange DNA. Another example of a trait that is likely to confer a strong selective advantage is the ability to fix atmospheric nitrogen gas, thereby obviating the need for sources of fixed nitrogen. With the possible exception of *Streptomyces thermoautotrophicus*, all nitrogen fixing bacteria use the same proteins and sequencing of the genes indicates that their base sequence is similar. When these genes moved among the different hosts is not known, but is probably many millions of years ago.

It would appear, therefore, that for bacteria any gene might be transferred to any bacterium, especially if the gene in question is plasmid-borne or bounded by insertion sequences. Genes located on chromosomes and not bordered by insertion sequences are less likely to be transferred and will usually require a host with a homologous region of DNA into which any transferred DNA might insert. The exception would be cases in which plasmids interacted with genomic DNA and picked-up genes to form plasmid-primes. In the laboratory it is reasonably simple to select for plasmid-primes by crossing bacteria containing certain types of plasmid with bacteria having non-homologous genomes, and selecting for inheritance of chromosomal genes (Johnston et al., 1978; Holloway, 1986). Such plasmids can be surprisingly stable and must occur naturally.

On the assumption that at least as much has yet to be understood about bacterial gene transfer as is now known, we might assume that further genes that facilitate widespread gene transfer might be found. We can also assume that while such genes may be widespread and may have facilitated horizontal gene transfer during the evolution of bacteria, their activities are sufficiently curtailed to enable well-defined species to develop. However, as we have only cultured less than 10% of the "species" of bacteria in soils, such comments should be considered in light of what we do not know. It is not inconceivable, although I believe it is unlikely, that most micro-organisms are constantly undergoing genetic alterations through naturally occurring gene transfer, but we never isolate them.

Whether horizontal gene transfer among bacteria will affect bacterial biodiversity or not is at present impossible to determine. Populations of microorganisms and the relative proportions of different bacteria are determined by so many different factors that it is not feasible to carry out the experiments to get useful information. We might expect from the ability of human pathogens to survive despite the extremely widespread use of antibiotics that adaptability is so great that gene transfer is likely to be the least of our concerns about microbial biodiversity.

Viruses

The most rapidly evolving genomes are those belonging to viruses. For some, such as RNA viruses, this is because mutation rates are very high, but for many it is because the selection pressures they are subject to have been affected strongly by human intervention. For example, for some diseases caused by viruses, such as measles, large populations of humans are required to maintain infection rates and such populations only arose within the last few thousand years. Likewise for some plant pathogenic viruses evolution has been rapid recently because it has only been in the last few decades that sufficiently large monocultures of host plants have existed to be able to allow strong selection on large virus populations.

Evidence for horizontal gene transfer comes from sequence studies which indicate that not only are viruses recombining (Aaziz and Tepfer, 1999), but also plant genes are being picked-up by viruses (Mayo & Jolly, 1991), and virus genomes are being picked-up by plants (Bejarano et al., 1996). The former is to be expected, as large populations of virus molecules are present in cells infected by more than one type of virus and recombination at regions of homology is likely. Some viruses are also able to indulge in a form of recombination arising during replication. So-called template switching occurs when replication of one virus nucleic acid switches to another virus adjacent to the starting strand, thereby forming a single molecule derived from two virus nucleic acids. It is clear from studies of viral nucleic acids that evolution is rapid and is driven in part by horizontal gene transfer. Whether or not the observation that viruses pick-up plant genes is reflected in the transfer of such genes among host plants remains to be demonstrated. It seems unlikely that this does not occur.

We do have evidence for the inheritance of viral DNA by cells in the intestines of rodents (Doerfler et al., 1998), and might assume that such cells are able to take-up other nucleic acids. However, for such events to be of significance in the evolution of animals there is a need for the inherited nucleic acid to be passed to progeny through germ cells, which is extremely unlikely. Indeed, inheritance of genes through gut cells needs to be a very rare event because all animal digestive systems are constantly exposed to the nucleic acids released from the plants, microbes, and animals they eat.

Sequence information for a virus that specifically infects insects (baculoviruses) showed that the enzyme (chitinase) it uses to digest insect hosts is almost certainly derived from the bacterium *Serratia marcescens* (Hawtin et al., 1995). How such a transfer might have occurred may never be determined.

Conclusions

It is clear from this brief survey of horizontal gene transfer that for bacteria it is probably very common. Evidence for the movement of genes across very different groups of organisms is coming primarily from sequence information. While we might be pleased with the number of species for which sequence information is available, it is salutary to remember just how many species have not been sequenced, and therefore how much horizontal gene transfer has yet to be identified.

For bacterial genes on plasmids under strong selection (such as with antibiotic therapy) it appears likely that horizontal gene transfer will be frequent and rapid, giving rise to plasmid variants in days or weeks. For genes that are under less strong selection, such as those for nitrogen fixation, the evidence is that while horizontal transfer can occur their benefit to recipient organisms is insufficient for nitrogen fixation to be a widespread property of bacteria. What determines rates of spread of genes, and thus how one might predict the spread of a particular gene, remains a mystery. Horizontal gene transfer has been an important feature in the evolution of species and will continue to be important, but there is no obvious reason why its rate should be enhanced by biotechnology.

References

Aaziz R, Tepfer M (1999) Recombination in RNA viruses and in virus-resistant transgenic plants. *Journal of General Virology* 80: 1339-1346

Andersson SGE, Kurland CG (1998) Ancient and recent horizontal transfer events: The origins of mitochondria. *Acta Pathologica, Microbiologica et Immunologica Scandinavica, Supplementum* 84 (106): 5-14

Bejarano ER, Khashoggi A, Witty M, Lichtenstein C (1996) Integration of multiple repeats of geminiviral DNA into the nuclear genome of tobacco during evolution. *Proceedings of the National Academy of Sciences of the USA* 93: 759-764

Buchanan-Wollaston AV, Passiatore JE, Cannon F (1987) The *mob* and *oriT* mobilization functions of bacterial plasmid promote its transfer to plants. *Nature* 328: 172-175

Buhariwalla H, Mithen R (1995) Cloning of a *Brassica* repetitive DNA element from resting spores of *Plasmodiophora brassicae*. *Physiological and Molecular Plant Pathology* 47: 95-101

Doerfler W, Schubert R, Heller H, Hertz J, Remus R, Schröer J, Kämmer C, Hilger-Eversheim K, Gerhardt U, Schmitz B et al. (1998) Foreign DNA in mammalian systems. *Acta Pathologica, Microbiologica et Immunologica Scandinavica, Supplementum* 84 (106): 62-68

Fraley RT, Rogers SG, Horsch RB (1986) Genetic transformation of higher plants. *CRC Critical Reviews of Plant Sciences* 4: 1-46

Fry JC, Day MJ (eds.) (1990) *Bacterial Genetics in Natural Environments*. 259 pages. Chapman and Hall, London

Hawtin RE, Arnold K, Ayres MD, Zanotto PM de A, Howard SC, Gooday GW, Chappell LH, Kitts PA, King LA, Possee RD (1995) Identification and preliminary characterisation of a chitinase gene in the *Autographica californica* nuclear polyhedrosis virus genome. *Virology* 212: 673-685

Holloway BW (1986) Chromosome mobilisation and genomic organisation in *Pseudomonas*. In: JR Sokatch (ed.) *The Bacteria* 10: 217-249. Academic Press, New York

Johnston AWB, Setchell SM, Beringer JE (1978) Interspecific crosses between *Rhizobium leguminosarum* and *R. meliloti*: formation of haploid recombinants and of R-primes. *Journal of General Microbiology* 104: 209-218

Lilley AK, Fry JC, Day MJ, Bailey MJ (1994) *In situ* transfer of exogenously isolated plasmid between indigenous donor and recipient *Pseudomonas spp*. In sugar beet rhizosphere. *Microbiology* 140: 27-33

Lin X, Kaul S, Rounsley S, Shea T, Benito MI, Town CD, Fujii CY, Mason T, Bowman CL, Barnstead M et al. (1999) Sequence and analysis of chromosome 2 of the plant *Arabidopsis thaliana*. *Nature* 402: 761-768

Mayer K, Schüller C, Wambutt R, Murphy G, Volckaert G, Pohl T, Düsterhöft A, Stiekema W, Entian KD, Terryn N et al. (1999) Sequence and analysis of chromosome 4 of the plant *Arabidopsis thaliana*. *Nature* 402: 769-777.

Mayo MA, Jolly CA (1991) The 5' terminal sequence of potato leafroll virus RNA: evidence of recombination between virus and host RNA. *Journal of General Virology* 72: 2591-2595

Maynard Smith J, Dowson CG, Spratt BG (1991) Localised sex in bacteria. *Science* 349: 29-31

Neilsen KM, Gebhard F, Smalla K, Bones AM, van Elsas JD (1997) Evaluation of possible horizontal gene transfer from transgenic plants to the soil bacterium *Acinetobacter calcoaceticus* BD413. *Theoretical and Applied Genetics* 95: 815-821

Raybould AF, Gray AJ (1993) Genetically modified crops and hybridisation with wild relatives: a UK perspective. *Journal of Applied Ecology* 30: 199-219

Rufener Al Mazyad P, Ammann K (1999) Biogeographical assay and natural gene flow. In: Ammann K, Jacot Y, Simonsen V, Kjellsson G (eds.) *Methods for Risk Assessment of Transgenic Plants, Volume III: Ecological risks and prospects of transgenic plants, where do we go from here? A dialogue between biotech industry and science*. Birkhäuser Verlag, Basel, 95-98

Rufener Al Mazyad P, Ammann K (1999) The *Medicago falcata/sativa* complex, wild relative introgression in Switzerland. In: LWD van Raamsdonk, den Nijs JCM. (eds.) *Plant Evolution in Man-Made Habitats*, Proceedings VIIth International IOPB Symposium, Amsterdam, 10-15 August 1998, Hugo de Vries Laboratory, Amsterdam, 271 -286

Touch A, Krieft S, Kalinowski J, Pühler A (2000) The 51,409-bp R-plasmid pTP10 from the multiresistant clinical isolate *Corynebacterium striatum* M82B is composed of DNA segments initially identified in soil bacteria and in plant, animal, and human pathogens. *Molecular and General Genetics* 263: 1-11

Methods for Risk Assessment of Transgenic Plants
IV. Biodiversity and Biotechnology
K. Ammann, Y. Jacot and R. Braun (eds)
© 2003 Birkhäuser Verlag Basel/Switzerland

Vertical gene flow

Klaus Ammann and Yolande Jacot

Botanical Garden, University of Bern, Altenbergrain 21, CH-3013 Bern, Switzerland

Introductory remarks

By comparing potential and actual risks in both fields of agriculture, with or without GMOs, we need a clear view of the similarities and differences. Raamsdonk (1993) has given a scheme of gene flow, whether natural or influenced by human activity. It clearly shows that most pathways of gene flow are both natural and also strongly influenced by agriculture and breeding. But there is one notable exception: Transformation is something quite new. It is now possible to jump over the limits of species, genera and even families and orders in a much easier way. Let's not forget that there is something universal about the sequences used for transformation and often lay persons are puzzled by expressions like "a fish gene" for instance. They have not the knowledge how to judge the novelty of such transformation. But this statement also counts for the experts: position effects and pleiotropic effects after transformation have to be considered in all cases, this is why we need to think about new risk assessment methods for genetic engineering. But we should always be aware of the fact that classical breeding is also dealing with the introduction of genes to a crop plant. The big difference is that these genes introduced in classical breeding are always belonging to some relatively close parents, transformation is always done in the relatively narrow environment of interbreeding species groups. But again we have to consider all cases in classical breeding, where single gene mutations or the change of only a small number of genes has dramatically changed morphology, growth and yield of a crop, especially in the case of maize and wheat. Considering the possible escape of transgenes to wild relatives there are parallels between classical breeding and genetic engineering regarding the processes of vertical gene flow, but the results may be different in the case of a possible escape of a novel transgene having new ecological potential.

Outcrossing on neighbouring natural area: gene transfer and cross-hybridization

Genetic engineering is obviously different and more powerful than traditional breeding. The question is whether this difference and the power translate into greater, more serious risks, or whether the risks are different from those of traditionally bred products. At this early stage in the use of the technology, it is impossible to answer this question with certainty.

Gene escape depends on many ecological and agronomical factors: whether or not the crop is allowed to flower; how far its pollen travels, success of fertilisation, extent of seed dispersal, seed survival and so on. Even if a gene does escape, its future may be bleak if it handicaps its new host (Young, 1989).

The probability of successful pollination depends on a great number of interrelated factors, such as: level of pollen production of (transgenic) plant, rate of self- and cross-fertilisation of receptor plant, rate of dispersion of donor pollen, properties of pollinating agents, spatial distance between pollen donor and wild recipient population, local density of recipient population, difference in phenology between crop and wild population.

Families with economically important crops often show low hybridization barriers. Simmonds (1976) gives an overview of the history of major and minor crop plants. It appears that hybridization was a common technique in the breeding history of 31 out of 39 families, and in 70 out of 90 genera, i.e. 75-80 percent of the cases.

Adaptative traits acquired under domestication usually have no selective value in natural habitats, and sustainable genetic introgression from domestic plants to wild relatives is rare. Gene exchange is more known in the other direction, from wild relatives to crops (Newmann, 1990).

There are more barriers to hybridization (Hadley and Openshaw, 1980):

a) External barriers such as spacial, ecological, phenological, mechanical and pollination ecological isolation phenomena.
b) Internal barriers such as prevention of fertilisation, hybrid weakness or inviability, hybrid sterility, hybrid breakdown in the F2 generation.
c) There are also barriers to pollen dissemination: Pollen viability, pollen sedimentation, distance of receiving plants, separation of anthers and stigmas by various contrivances.

Sexual reproduction occurs among crops, weeds, and wild plants within the distance that pollination can occur and where the plants are sexually compatible. Because pollen transfer is mediated primarily by wind and insects, the distance within which pollination can occur is affected by wind turbulence, speed and direction and/or the flying range of insects (Regal, 1982). Another factor is the longevity of the pollen itself. Generally, pollen is viable only a short time ; that is, it is capable of fertilising eggs only within a time period of minutes, hours or days after being produced.

Of particular importance is the aerial transport of pollen grains. With plants that are pollinated by insects, one approach is to follow the pollinator. Insects such as honeybees and bumblebees are often loyal to a particular variety of flower for a given period of time. Moreover, they prefer short trips between blooms, visiting several flowers on a single plant perhaps, and then flying to a near neighbour, a few meters away at most. Yet because bees do not deposit all the pollen from one plant to the next, genes tend to travel further than appearances suggest (genes travelled twice that distance between two stops of bees) (Young, 1989). With plants that are pollinated by wind, most pollen is deposited near to home, but a small amount travels much farther (Young, 1989). It should also taken into account that indigenous pollinatior insects often are better adapted to local variation in climate than honey bees (Boyle-Makowski et al., 1985).

Do transgenes have an influence in hybridizing dynamics of crops?

Theoretically, genetic modification may affect hybrid formation either by changing frequency with which it occurs, or by altering the range of species with which the crop is sexually compatible. Increasingly the evidence suggests that modification has little or no impact on either factor with the present day transgenes released in the fields. In general, plant fertility is maintained after the genetic modification, although there are instances where a particular transformation protocol has produced sterile plants. For example, transgenic maize regenerated from protoplasts is sterile if the protoplasts are transformed by electroporation (Rhodes et al., 1988), but fertile if transformed by polyethylene glycol (Omirulleh et al., 1993). Fertility problems in transgenic plants may therefore be due to particular combinations within the transformation protocol and starting material rather than due to the transformation in itself. It is unlikely that totally sterile transgenics would be of interest as they are unsuitable for breeding purposes. It may be assumed, therefore, that commercial transgenic crop varieties will show similar fertility as non-transgenics. One exception is the crop modified for male sterility. These crops are useful to plant breeders.

One of the most important factors of hybridization is the movement of the fertile pollen to a receptive stigma. Clearly, pollen production in the crop and the stigma-ripening in the wild relative must be concurrent. If the phenology of a transgenic crop is different from the unmodified form, then the potential for hybridization will be modified. Several genetic modifications have produced alterations in flower development, but these tend to be changes in structure, leading to morphologically abnormal flowers (Mizukami and Ma) rather than the production of normal flowers with different flowering periods. These experiments were done with genes known to be involved in flower development or promoters only activated in floral tissue. Up to now, transformation does not seem to alter the flowering period. However, as flowering period is a multigenic trait there is a possibility that a transgene may integrate into and disrupt the expression of a gene involved in flowering-period determination. Assuming simultaneous flowering, the extent of pollen transport will determine the potential for crop pollen to land on the stigma of a wild relative. Pollen flow, in most cases, describes a highly leptokurtic distribution from the source plant, with most grains moving less than 2 m in herbaceous plants (Levin and Kerster, 1974). Recent work confirms this for transgenic pollen (Umbeck et al., 1991). However, it is misleading to conclude from these data that hybridization between crops and wild relatives can only occur over short distances. Even in case a hybridization over long distances takes place only rarely, it may produce a new, viable founder population.

Pollination is a major determinant factor for the hybridization rate, but the types of hybrids are determined by interspecific incompatibility mechanisms. The genetics of these mechanisms are still not clear. Unless interspecific incompatibility is controlled by one or a small number of genes, the transformation process is unlikely to alter the cross-compatibility of the plants following transformation. Transformation of a self-compatible species with an S-allele involved in self-incompatibility mechanisms has been achieved, but the transgenic plants became self-compatible (Shiba et al., 2001). Introduction of an incompatibility system into self compatible species will be useful to plant breeders, but it is not clear whether self incompatibility is involved in interspecific incompatibility, and, consequently, such modifications can be expected to have little effect on interspecific hybridization.

Another important question is whether or not the relation between pollinators and plants will be modified by transformation. First the new proteins synthesized by the transgenic plant should be toxic for bees. The non-toxicity of the chitinase has been proved (Pham-Delègue et al., 1992). Could the character of a transgenic plant modify the activity of the insect? The same group demonstrated that the foraging time was shorter on a transgenic plant than on a non-transgenic one. However, Pham-Delègue et al. concluded that transgenic oilseed rape has no negative effects on foraging bees under controlled conditions. Similarly, Paul et al. (1991) found that there are no differences in the range of animals and the frequency of visits between modified and non-modified tobacco plants.

Insects like bees are attracted by light of a wave length between 300 nm and 650 nm (Dumas, 1984). This includes part of the ultraviolet light but excludes the long-waved red light. A change in the flower colour could disturb the attractiveness of the flower for the insects and change cross-pollination rate for insect-pollinated plants.

On the other hand, genetic engineering may be favourable to the hybridization process. Studies regarding alfalfa *(Medicago sativa)* treated by a pesticide (dimethoate) clearly demonstrated that this pesticide is found in the pollen and the nectar at a very low level but even at a low level this pesticide is toxic for the bees (Dumas, 1984). This has an effect on the entomophilous pollination. Plants resistant to diseases will be treated with less pesticide, so that the efficiency of pollination for entomophilous plants could be enhanced.

The assessment of a potential risk regarding field release of transgenic crops: introduction to gene flow indices

As a result of discussions in the symposium at Louverain, we propose gene flow indices after the idea of some Dutch authors ([Frietema] Frietema De Vries, 1996). We are giving here an adapted version in order to spur discussion on an European level. We think it is desirable to establish a European classification system as proposed by Frietema De Vries (1996), where some of our proposals have been adopted. It is not possible to arbitrate the crops and their wild relatives on one and the same level all over Europe: Classification work has to be done on a regional scale taking into account local environmental conditions, species and transgenes. This regional scale has been proposed by Frietema De Vries (1996), following the well known subdivisions of Meusel.

Here we deal only with the three first codes, but we feel strongly the necessity of a fourth code for the future: We need to assess also the effect of the inserted transgene itself. For this Dg code we need experimental approaches on all levels from a strict containment over small scale field releases to the large scale releases over longer periods. For the time being there remains only the possibility of a rough estimate of how transgenes will have side effects in the long run, some comments are built in provisionally in code Dp (vertical gene flow). The authors are well aware of the pragmatic view they take, which is blurring the logic of the three codes already defined.

These codes are presented here in order to open debate on feasibility and organisation of such codes for future risk assessment: The codes can serve as a first rough estimate, before going into more detail for a risk assessment based on field monitoring and experimental approach, where judged necessary.

Dpdf-codes, adapted to Swiss and European needs as a whole

Classification of the codes of dispersal of pollen (Dp)

Dispersal of pollen and hybridization potential, including a differentiation of possible negative ecological effects of the inserted gene itself. Categories 0 (lowest risk) to 5 (highest risk) and U (unknown).

Category Dp 0

No chance for hybridization because there are no wild relatives growing in Switzerland. No ecological effects when the cultivated plants come into flower.

Monitored field releases possible, no containment experiments and no field experiments necessary.

Category Dp 1

No chance for hybridization with wild relatives because it is experimentally proven that wild species of the same genus in Switzerland are not compatible with the cultivated plant: (artificial pollination methods and/or embryo rescue are necessary to produce hybrids).

No ecological effects when cultivated plants come into flower. Monitored field releases possible without containment. However, experiments should be carried out, to test there are no negative effects on the host/predator system in case of transgenes introducing new resistance and/or competition effects.

Category Dp 2
No chance for hybridization with wild relatives because there is no record of spontaneously formed hybrids of the cultivated plant with wild species of the same genus in Switzerland.

However, hybridization is possible under experimental conditions and progeny is fertile without any artificial help. Chances of gene flow by hybridization is small due to various outcrossing barriers (competition, biogeographical or ecological incompatibility), but under special local or artificial conditions in agricultural systems still to be considered as possible rare events.

a) In certain species groups there is a small chance of getting new transgenic hybrids, but no invasions are to be expected.
b) In other species groups there is a small chance of getting new transgenic weeds which tend to be aggressive and will possibly cause invasions under unfavourable conditions.

Category Dp 3
Natural hybridization occurs only occasionally, backcrosses have not been observed up to now. Local situations have to be studied carefully in risk assessment of field experiments. Species to species, region by region and step by step approach required.

a) In certain species groups and under unfavourable circumstances gene flow by pollen transfer will occur, but new transgenic hybrids do not tend to be invasive.
b) In other species groups and under unfavourable circumstances gene flow by pollen transfer can influence ecosystems negatively: Local invasions of new transgenic weeds will occur.

Category Dp 4
Chance for natural hybridization is medium; backcrosses have been observed, successful outcrossing occurs fairly often. Natural fertile hybrids are sometimes observed, small hybrid populations can be detected in nature. Species to species, region by region and step by step approach required.

a) Transgenic hybrids will have no ecological effects on the flora of the Switzerland, since the new hybrid is only capable to invade small ecological niches, and therefore does not demonstrate any disturbing invasiveness, since the inserted gene itself did not show negative ecological effects in long term monitoring experiments. Experiments should also be carried through proving that there are no negative effects on the host / predator system.
b) Transgenic hybrids will have ecological effects on the flora of the Switzerland, since the new weed is capable to invade ecological niches, and therefore is potentially demonstrating invasiveness. There may also be negative effects (e.g. more competitive, more allelopathic) caused by the inserted gene itself.

Category Dp 5
Chance for natural hybridization is high; vertical gene flow occurs often, hybrids are fertile and backcross frequently. Hybrid populations are often found in nature. Species to species, region by region and step by step approach required.

a) Transgenic weeds will have no ecological effects on the flora of Switzerland, nevertheless the new weed is capable to invade important ecological niches and it will act as a new weed (which should by all means be avoided!), but the inserted gene itself does not show negative ecological effects.
b) Transgenic weeds will have negative ecological effects on the flora of Switzerland since it is capable of invading many ecological niches as a major new weed and/or since the inserted gene itself may have characters demonstrating negative ecological effects.

Category Dp U:
Data too scanty or lacking at all, no evaluation possible.

Classification of the codes for the dispersal of diaspores (Dd)

Category Dd 0
No chance for dispersal of diaspores to the wild: Seeds are sterile or otherwise deficient, they have lost reproductive function. No ecological effects are expected from fruiting of the cultivated plants.

Category Dd 1
Dd to the wild occurs only occasionally and under very favourable conditions, plants usually survive only for one season (advena), they are not adapted for survival in our climate. No ecological effects are to be expected regarding the Swiss ecosystem.

Category Dd 2
Chance for dispersal of diaspores to the wild is small, but under favourable and exceptional conditions possible. Further research on population dynamics seems necessary. For risk assessment the standing of the plants in the Swiss ecosystem can be of importance.

Category Dd 3
Chance for dispersal of diaspores (by spontaneous vegetative reproduction) is real; fruiting of the cultivated plant is essentially undesirable and will normally be suppressed by various methods. Further research on population dynamics is necessary. For risk assessment the standing of the plants in the Swiss ecosystem can be of importance.

Category Dd 4
Chance for dispersal of diaspores to the wild real. Fruiting of the cultivated plant occurs normally during cultivation. Ecological effects can be expected from fruiting of the cultivated plant. For risk assessment the standing of the plants in the Swiss ecosystem will be of importance.

Category Dd 5
Dispersal of diaspores to the wild will be the rule. Fruiting occurs very frequently and also extremely abundant. Ecological effects can be expected from fruiting of cultivated plant. For risk assessment the standing of the plants in the Swiss ecosystem will be of importance.

Category Dd U
Data too scanty or lacking at all, no evaluation possible.

Classification of the codes for Df (frequency of distribution)

Category Df 0
No plants of this species or of a wild relative, no feral populations found in nature; no ecological effects are expected from the introduction of the cultivated transgenic plant.

Category Df 1
Plants of this species or of wild relatives are extremely rare in the wild and have their stable place in the Swiss ecosystem in specific associations. No feral populations are found in Switzerland. Chances for hybridizing with the wild or feral populations are negligible. Locations to grow transgenic plants should be appropriately chosen in order to avoid hybridization and any ecological effect.

Category Df 2
Plants of this species or of wild relatives are rare, but occur sporadically, distribution difficult to predict and essentially uncontrollable. Feral populations may exist in certain regions. Chances for hybridizing with wild populations are scanty but unpredictable. Ecological effects from the introduction of the cultivated plant may be expected, but in most cases on a local scale only. Locations to grow transgenic plants should be appropriately chosen in order to avoid hybridization and any ecological effect.

Category Df 3
Plants of this species or of wild relatives are not very common in the wild and have their stable place in Swiss ecosystem. Feral populations are known from Switzerland, but not frequent. Chances of hybridizing with the wild populations exist but are small. Some ecological effect from the introduction of the cultivated plant may be expected under unfavourable conditions when cultivated plants and wild relatives are not sufficiently separated. Locations to grow transgenic plants should be carefully chosen in order to avoid hybridization and any ecological effect.

Category Df 4
Plants of this species and their wild relatives are not frequent but well distributed over the whole Swiss plateau, chances for hybridizing with wild populations are considerable, but under very favourable conditions it can still be safely prevented. Feral populations are known and distributed over an important part of Switzerland. Locations to grow transgenic plants should be carefully chosen in order to avoid hybridization and any ecological effect. Detailed biogeographical studies are necessary to reach this goal.

Category Df 5
Plants of this species and their wild relatives are common and well distributed over the whole Swiss plateau, chances for hybridizing with wild populations must be expected and cannot be prevented in field experiments. Feral populations are frequent and distributed over the whole Switzerland. In exceptional cases locations to grow transgenic plants can still be carefully chosen in order to avoid hybridization and any ecological effect. Detailed biogeographical studies are necessary to reach this goal

Category Df U
Data too scanty or lacking at all, no evaluation possible.

Vertical gene flow classification by combination of the three codes

The goal of this study was to develop a convenient classification of gene dispersal probability from transgenic crop to the wild flora, adapted for Switzerland. After an evaluation of the three single factors (see above, dispersal codes), the combination of these codes enables us to estimate impact on the wild flora. Five categories of risk probability have been developed:

After an evaluation of the three single factors, their combination enables us to estimate the impact of a transgenic species on the environment. Six categories of risk probability have been developed:

(1) No effect
No related species or no compatible related species of the crop are known in Switzerland. Field releases of species belonging to this category are possible without any containment or short term monitoring. Certain transgenes have to be tested in medium term field experiments regarding their secondary effects on ecosystems: Sustainable resistance must be achieved. To reach this goal a long term monitoring is required.

(2) Minimal effects
No records of spontaneous hybridization between the crop and the wild relatives are known in Switzerland. Field releases are possible after a thorough clarification of the biogeographical situation. Short term monitoring in confinements should be done prior to large scale field releases. Certain transgenes have to be tested in medium term field experiments regarding their secondary effects on ecosystems (pest and insect resistance genes).

(3) Low but local effects
Gene flow occurs towards wild or feral species existing also outside agricultural environment and control. Release experiments should first be done in confinements and afterwards in small scale releases closely monitored. This statement is restricted to transgenes not causing enhanced competitiveness outside agricultural environment, such as herbicide tolerance. Any other transgenes should be carefully tested in confinements.

(4) Substantial but local effects
Gene flow is high and substantial, but still locally controllable. Field releases could be done within strict confinements. A case by case analysis including the potential effects of the transgene is required before any field releases are done. Long term monitoring of field releases under strict biological or geographical confinement conditions is necessary in order to study competitiveness of the transgenic crop. Risky transgenes have to be avoided.

(5) Substantial and wide-spread effects
Gene flow is high, substantial, and widespread and will not be controllable by any means. No field releases of species belonging to this fifth category are possible. Medium term monitoring under strict confinement conditions is necessary in order to find out about competitiveness of the transgenic varieties. Experiments with less risky crop varieties (e.g. with male sterility) having the same favourable effect desired.

(6) Unknown (one of the three indices is unknown)
More studies are needed before any field releases are done.

Conclusions

Besides direct measurement of transgene flow to wild relatives we need a system which builds on decades old experience. This is why we think that the most reliable data do not stem from short term field experiments of outcrossing schemes, which are always dependent on meteorological conditions and regional characteristics. This decades old data material is obtainable

Table 1: Risk categories for some crops important to Switzerland

		Dp code					
Df code	**Dd code**	**0**	**1**	**2**	**3**	**4**	**5**
0	0						
	1	Tomato					
	2	Tobacco					
	3						
	4						
	5						
1	0						
	1						
	2		Beet				
	3						
	4						
	5						
2	0						
	1						
	2				Endive		
	3						
	4				Turnip		
	5						Lettuce
3	0						
	1						
	2				Cabbage		
	3				Radish		
	4						
	5				Rape		
4	0	Maize					
	1			Barley			
	2			Wheat		Carrot	
	3			Rye	Chicory		
	4						
	5						
5	0						
	1	Potato					
	2						
	3		Clovers				
	4						Alfalfa
	5						Grasses

No effect
Minimal effect
Low but local effect
Substantial but local effect
Substantial and wide-spread effect

in herbaria, which always contain a rich material on hybrids, collected by specialized taxonomists. Taxonomists have a very sharp eye for unusual morphology of hybrids within large populations, and usually, hybrid specimens are over-represented in well assorted big herbaria, with collectors items gathered over centuries. Morphometrics add to the precision of hybrid recognition in herbaria, allow or statistically sound data sets and give a fairly well documented picture on hybrid processes out in the field, which is at least as valuable a data set compared to lab, garden and short term field experiments. Our Dutch-Swiss method of recognizing hybrids with other means than experimental ones needs to be enhanced in the view of biogeography. It would be necessary to regionalize the Dp codes in Europe, since the biogeography of the wild relatives for many crops vary considerably.

References

Abbott R (1994a) Ecological risks of transgenic crops. *Tree* 9: 280-281

Abbott R (1994b) Reply to Thacker JRM. Transgenic plants. *Tree* 9: 486

Adams WT, Griffin AR, Moran GF (1992) Using paternity analysis to measure effective pollen dispersal in plant populations. *The American Naturalist* 140 (5): 762-780

Ahl Goy P, Chasseray E, Duesing J (1994) Field trials of transgenic plants: an overview. *Agro-Industry hi-tech* March/April, 10-15

Ahl Goy P, Duesing J (1995) From Plots to Plots: Genetically Modified Plants on Trial. *Biotechnology* 13: 454-458

Ahl Goy P, Duesing J (1996) Assessing the Environmental Impact of Gene Transfer to Wild Relatives. *Biotechnology* 14: 39-40

Altmann M (1994) Schlussfolgerungen aus dem Technikfolgenabschätzungs-Verfahren (TA) zum Anbau von Kulturpflanzen mit gentechnisch erzeugter Herbizidresistenz. *Gaia*, 3(6)

Altmann M and Ammann K (1992) Gentechnologie im gesellschaftlichen Spannungsfeld: Züchtung transgener Kulturpflanzen. *Gaia*, 1(4): 204-213

Ammann K (1995) Gentechnisch veränderte Kulturpflanzen in der Umwelt: Chancen und Probleme – Die Suche nach gangbaren Wegen. In: Erbersdobler HF, Hammes W, Jany KD (eds.): *Gentechnik und Ernährung.* Wissenschaftliche Verlagsgesellschaft mbH Stuttgart, 161-195

Ammann K (1995) Die ökologischen Risiken der Gentechnologie und wie wir damit umgehen können. In: Behrens, M, Meyer-Strumberg S, Simonis G (Hrsg), *Gentechnik und Nahrungsmittelindustrie.* Mensch undTechnik, Sozialverträgliche Technikgestaltung. p.177-190

Ammann K, Felber F, Keller J, Jacot Y, Küpfer Ph, Rufener P, Savova D (1994) Dynamic biogeography and natural hybridization of selected weedy species in Switzerland. Symposium "Gene Transfer: Are Wild Species in Danger?", Le Louverain, Switzerland, Nov.1994. *Environmental Documentation*, Federal Office of Environment, Forests and Landscape (FOEFL), 12: 36-39

Anderson C (1992) Researchers ask for help to save key biopesticide. News. *Nature* 355: 661

Anderson E (ed.) (1949) *Introgressive hybridisation.* John Wiley & Sons, Inc., New York, Chapman & Hall, Limited, London

Andow DA (1994) Community response to transgenic plant release: using mathematical theory to predict effects of transgenic plants. *Molecular Ecology* 3: 65-70

Arnold ML and Hodges SA (1995) Are natural hybrids fit or unfit relative to their parents ? *Tree.* 10 (2): 67-71

Ashman TL, Galloway LF, Stanton ML (1993) Apparent vs. effective mating in an experimental population of *Raphanus sativus. Oecologia* 96: 102-107

Austin RB, Flavell RB, Henson IE, Lowe HJB (eds.) (1986) *Molecular biology and crop improvement.* Cambridge University Press

Bainton JA (1993) Considerations for release of herbicide resistant crops - Volunteer crops as weeds. *Aspects of Applied Biology* 35: 45-52

Baker HG (1967) Characteristics and modes of origin of weeds. In : HG Baker, Stebbins GL (eds.) *The genetics of colonizing species.* Academic Press, New York and London, 147-172

Baker HG (1974) The evolution of weeds. *Ann. Rev. of Ecology and Systematics* 5: 1-24

Barrett SCH (1983) Crop mimicry in weeds. *Econ. Bot.*, 37 (3): 255-282

Bartsch D (1995) Ökologische Begleitforschung zu rizomaniaresistenten Zuckerrüben. In: Albrecht S, Beusmann V (eds.): *Ökologie transgener Nutzpflanzen, Gentechnologie - Chancen und Risiken.* Campus Verlag, Frankfurt/New York, 31: 81-98

Bartsch D, Sukopp H, Sukopp U (1993) Introduction of plants with special regard to cultigens running wild. In: Wöhrmann K, Tomiuk J (eds.): *Transgenic organisms*. Birkhäuser Verlag Basel/Switzerland, 135-151

Bazin MJ and Lynch JM (eds.) (1995) *Environmental Gene Release. Models, Experiments and Risk Assessment*. Chapman & Hall, New York, 166p

Becker T (1951) Siebenjährige blütenbiologische Studien an den Cruziferen *Brassica napus* L., *Brassica rapa* L., *Brassica oleracea* L., *Raphanus* L. und *Sinapis* L.. Teil I. *Zeitschrift für Pflanzenzüchtung*, 29: 222-240

Bevan MW, Harrison BD, Leaver CJ (eds.) (1995) *The Production and Uses of Genetically Transformed Plants*, Chapman & Hall, New York

Beversdorf WD, Scott LS (1987) Development of Triazine Resistance in Crops by Classic Plant Breeding. *Weed Science* 35: 9-11

Blanck KD (1991) Raps in der Fruchtfolge nicht überziehen! In: Rapsanbau für Könner - Bestände gezielt aufbauen und schützen. Top agrar - *Das Magazin für moderne Landwirtschaft*. Landwirtschaftsverlag Münster-Hiltrup, 94-98

Bottermann J, Leemans J (1988) Engineering of Herbicide Resistance in Plants. *Biotechnology and Genetic Engineering Reviews* 6: 321-341

Boudry P, Mörchen M, Samitov-Laprade P, Vernet P, Van Dijk H (1993) The origin and evolution of weed beets: consequences for the breeding and release of herbicide-resistant transgenic sugar beets. *Theor. Appl. Genet.*, 87: 471-478

Boyle-Makowski RMD (1985) Pollinator activity and abiotic factors in an apple orchard. *The Canadian Entomologist* 117: 1509-1521

Crawley MJ (1992) The comparative ecology of transgenic and conventional crops. In: R Casper, Landsmann J (eds.): *2nd international symposium on the biosafety results of field tests of genetically modified plants and microorganisms*. May 11-14, 1992, Goslar, Germany, 43-52

Crawley MJ, Halls RS, Rees M, Kohn D, Buxton J (1993) Ecology of transgenic oilseed rape in natural habitats. *Nature*, 363 (6430): 620-623

Curtis H (ed.) (1983) *Biology*. New York : Worth Publishers, Inc.

Cussans GW (1978) The problem of volunteer crops and some possible means of their control. Proc. British Crop Protection Conference. *Weeds*, 3: 915-921

Dale P (1992) Spread of Engineered Genes to Wild Relatives. *Plant Physiol.* 100, 13-15

Dale PJ, Parkinson R, Scheffler JA (1993) Dispersal of genes by pollen - the PROSAMO-Project. *British Crop Protection Council Monograph* 55: 133-143

Dumas C (1984) Ecologie florale et pollinisation. In : Pesson P, Louveaux J (eds.). *Pollinisation et production végétales*. INRA, Paris, 31-46

Eckert JE (1933) The flight range of honeybee. *J. Agric.* Res. 47: 257-285

Eijlander R (1989) Transgene biet, ja of nee. *Prophyta*, 43: 232-233

Ellstrand N (1988) Pollen as a vehicle for the escape of engineered genes ? *Trends in Ecology and Evolution* 3: S30-S32

Ellstrand N (1992) Gene flow by pollen : implications for plant conservation genetics. *Oikos*, 63: 77-86.

Ellstrand N, Marshall D (1985) Interpopulation gene flow by pollen in wild radish, *Raphanus sativus*. *American Naturalist* 126: 606-616

Ellstrand N, Devlin B, Marshall D (1989) Gene flow by pollen into small populations : data from experimental and natural stands of wild radish. *Proceedings of the National Academy of Sciences* 86: 9044-9047

Feitelson J, Payne J, Kim L (1992) *Bacillus thuringiensis*: insects and beyond. *Bio/Technology* 10: 438-442

Ferre J, Real MD, Van Rie J, Jansens S, Peferoen M (1991) Resistance to the Bacillus thuringiensis bioinsecticide in a field population of *Plutella xylostella* is due to a change in a midgut membrane receptor. *Proc. Nat. Acad. Sci. USA* 88: 5119-5123

Fitter A, Perrins J, Williamson M (1990) Weed probability challenged. *Bio/Technology* 8: 473

Fredshavn JR, Poulsen GS (1995) Competitiveness of transgenic oilseed rape. *Transgenic Research* 4: 142-148

Friedt W (1995) Oral presentation 'Erwünschte und unerwünschte Inhaltsstoffe - eine Herausforderung für den Pflanzenbau' at the annual meeting 1995 of the Swiss Society of Agronomy in Bern, Switzerland

Frietema De Vries FT (1996) *Cultivated plants and the wild flora. Effect analysis by dispersal codes*. Thesis. Rijksherbarium, Hortus Botanicus, Leiden, 100p

Frietema De Vries FT, Van der Meidjen R, Bradenburg WA (1994) Botanical files on lettuce (*Lactuca sativa* L.) for gene flow between wild and cultivated lettuce (*Lactuca sativa* L. Including *L. Serriola* L., Compositae) and the generalized implications for risk assessment on Genetically Modified Plants. *Gorteria* suppl. 2, 44p

Gabriel W (1993) Technologically modified genes in natural populations: some skeptical remarks on risk assessment from the view of population genetics. In : Wöhrmann K, Tomiuk J (eds.): *Transgenic Organisms* Birkhäuser, Basel, 109-116

Graner A, Wenzel G (1993) Chancen und Risiken der Grünen Gentechnik. *Pflanzenschutz-Praxis* 4: 20-22

Grant V (ed.) (1981) Plant speciation. Second edition. New York, Columbia University Press

Gray A (1986) Green Industry Biotechnology Platform 1995. *Extracts from GIBiP Database on Field Trials 1986 - 1994*, G95/129

Gregorius HR, Steiner W (1993) Gene transfer in plants as a potential agent of introgression. In: K Wöhrmann, Tomiuk J (eds.), *Transgenic Organisms*. Birkhäuser, Basel 1993, 109-116

Gressel J, Regev Y, Malkin S, Kleifeld Y (1983) Characterization of an s-triazine-resistant biotype of *Brachypodium distachyon*. *Weed Science* 32: 450-456

Gressel J, Kleifeld Y (1994) Can wild species become problem weeds because of herbicide resistance? *Brachypodium distachyon*: a case study. *Crop protection* 13 (8): 563-566

Hadley HH, Openshaw SJ (1980) Interspecific and intergeneric hybridization. In: Fehr WR, Hadley HH (eds.). *Hybridization of crop plants*. American Society of Agronomy-Crop Science Society of America, Madison, 133-159

Halliun d' K, Sottermann J, Greef de W (1990) Engineering of herbicide resistant alfalfa and evaluation under field conditions. *Crop Science* 30: 866-871

Halluin D' K, Bossut M, Bonne E, Mazur B, Leemans J, Botterman J (1992) Transformation of sugarbeet (*Beta vulgaris* L.) and evaluation of herbicide resistance in transgenic plants. *Bio/Technology* 10: 309-314

Harding K, Harris PS (1994) *Risk assessment of the release of genetically modified plants : a review*. Ministry of Agriculture, Fisheries and Food (ed.), London, 54pp

Harlan JR (1965) The possible role of weed races in the evolution of cultivated plants. *Euphytica* 14: 173-176

Harlan JR, De Weet JMJ, Price EG (1973) Comparative evolution of cereals. *Evolution* 27: 311-325

Hartmann E, Schuldes H, Kübler R, Konold W (eds.) (1994) *Neophyten - Biologie, Verbreitung, Kontrolle ausgewählter Arten*. Ecomed. Landsberg, Germany, 302p

Herrmann G (1987) *Biochemische, physiologische und immunologische Untersuchungen zum Herbizid-Bindeprotein aus Thylakoidmembranen von einzelligen Algen und höheren Pflanzen*. Dissertation Universität Konstanz. 1.Aufl.

Heywood V and Zohary D (1995) A Catalogue of the Wild Relatives of Cultivated Plants Native to Europe. *Flora Mediterranea* 5: 375-415

Hojland JG, Pedersen S (eds.) (1994) *Carrot (Daucus carota L. ssp. sativus (Hoffm.) Arcangeli) : Dispersal, establishement and interactions with the environment*. The National Forest and Nature Agency, Copenhagen, Denmark. 67p

Holm L, Doll J, Holm E, Pancho J, and Herberger J (eds.) (1977) *The world's worst weeds. Distribution and biology*. John Wiley & Sons, Inc.

Holzner W (1982) Concepts, categories and characteristics of weeds. In: Holzner W, Numata N (eds.): *Biology and ecology of weeds*, Dr. W. Junk Publishers, The Hague, 3-20

Hoyle R (1993) Herbicide-resistant crops are no conspiracy. *BIO/TECHNOLOGY* ,1: 783-78

Jacot Y and Jacot P (1994) Application du génie génétique à l'agriculture : Evaluation des dangers potentiels pour la flore suisse. *Cahier de l'Environnement, Organismes* 235, 77p

Jörgensen RB, Andersen B (1994) Spontaneous hybridization between oilseed rape (*Brassica napus*) and weedy *B. campestris [B. rapa subsp. campestris]* (*Brassicaceae*): a risk of growing genetically modified oilseed rape. *American Journal of Botany* 81(12): 1620-1626

Keeler K, Turner C (1991) Management of Transgenic Plants in the Environment. In: M Levin, Strauss H (eds.): *Risk Assessment in Genetic Engineering*. Mc Graw Hill, 189-218

Kirkpatrick KJ, Wilson HD (1988) Interspecific geneflow in Cucurbita : C. texana vs. C. pepo. *Amer. J. Bot.* 75: 519-527

Kjellsson G, Simonsen V (eds.) (1994) *Methods for risk assessment of transgenic plants*. I. Competition, establishment and ecosystem effects. Birkhäuser, Basel

Kjellsson G, Simonsen V, Ammann K (eds.) (1997) *Methods for risk assessment of transgenic plants*. II. Pollination, Gene-transfer and Population impact. Birkhäuser, Basel

Lambelet-Haueter C (1990) Mauvaises herbes et flore anthropogène. I. Définitions, concepts et caractéristiques écologiques. *Saussurea* 21: 47-73

Lambelet-Haueter C (1991) Mauvaises herbes et flore anthropogène. II. Classifications et catégories. *Saussurea* 22: 49-81

Lamberti F, Waller JM, Van der Graaff NA (eds.) (1982) *Durable Resistance in Crops*. Plenum Press New York and London

Langevin SA, Clay K, Grace JB (1990). The incidence and effects of hybridisation between cultivated rice and its related weed red rice (*Oryza sativa* L.). *Evolution*, 44: 1000-1108

Leisinger K (ed.) (1991) *Gentechnik für die Dritte Welt ?* Birkhäuser, Basel

Levin D (1978) The origin of isolating mechanisms in flowering plants. In: M Hecht, Steere W, Wallace B (eds.): *Evolutionary Biology*. Plenum Press, New York, 185-317

Levin DA, Kerster HW (1974) Gene flow in seed plants. *Evol. Biol.* 7: 139-220

Lupi C (1995) Genetic engineering for plant protection - Methods, state of the art and applications. Technology assessment of pest resistant transgenic plants, report 1. BATS, 60p

Madsen KH (1994) Weed management and impact on ecology of growing glyphosate tolerant sugarbeets (*Beta vulgaris* L.). Ph.D.-Thesis. The Royal Veterinary and Agricultural University, Weed Science, Copenhagen, Denmark and MARIBO Seed, Holeby, Denmark, 61p

Mangelsdorf PC (1961) Introgression in Maize. *Euphytica* 10: 157-168

McGaughey WH (1985) Insect resistance to the biological insecticide *Bacillus thuringiensis*. *Science* 229: 193-194

McGaughey WH, William ME, Wilson E (1992) Managing Insect Resistance to *Bacillus thuringiensis* Toxins. *Science* 258: 1451-1455

Meer Van der PJ (eds.) (1993) *Proceedings (working document) Pan-European conference on the potential long-term ecological impacts of the release of genetically modified organisms.* Council of Europe, Strasbourg

Mikkelsen TR, Andersen B, Jörgensen RB (1996) The risk of crop transgene spread. *Nature* 380: 7

Mizukami Y, Ma H (1992) Ectopic expression of the floral homeotic gene AGAMOUS in transgenic Arabidopsis plants alters floral organ identity. *Cell* 71: 119-131

Newmann SM (1990) Pollen transfer from transgenic plants: Concepts for risks assessment and data requirements. *AAAS/EPA Environmental Science and Engineering Fellow* 1-40

NRC Report on Field Testing (1989) National Research Council of the National Academy of Sciences. Committee on scientific evaluation of the introduction of genetically modified micro-organisms and plants into the environment. *Field testing genetically modified organisms: Framework for decisions.* National Academy Press, Washington, D.C., 170 p

OECD (1993a) Field releases of transgenic plants, 1986-1992 an analysis. OECD, Paris, 39p

OECD (1993b) Safety considerations for biotechnology: Scale-up of crop plants. OECD, Paris, 41p

Oka HI, Morishima H (1982) Ecological genetics and the evolution of weeds. In: W Holzner, Numata, N (eds.): *Biology and ecology of weeds.* Dr. W. Junk Publishers, The Hague, 73-89

Omirulleh S, Abrahaqm M, Golovkin M, Stefanov I, Karabaev MK, Mustardy L, Morocz S, Dudits D (1993) Activity of a chimeric promoter with the doubled CaMV 35S enhancer element in protoplast-derived cells and transgenic plants in maize. *Plant Mol. Biol.* 21: 415-428

Oxtoby E, Huges MA (1990) Engineering herbicide tolerance into crops. *TIBTECH* 3/8: 61-65

Paul EM, Lewis GB, Dunwell JM (1991) The pollination of genetically modified plants. *Acta Horticulturae* 288: 425-429

Pham-Délègue MH, Picard-Nizou AL, Arnorld G, Grison R, Toppan A, Olsen L, Masson C (1992) Impact of genetically modified rapeseed on insect pollinators (honeybees). In : R Casper, Landsmann J (eds.): *Proceedings of the 2nd International Symposium on „the biosafety results of field tests of genetically modified plants and micro-organisms"* May 11-14, 1992, Goslar, Germany

Pickersgill B (1981) Biosystematics of crop-weed complexes. *Kulturpflanze* XXIX: 377-388

Poulsen GS, Fredshavn J (1995) Assessing competitive ability of transgenic oilseed rape. Workshop on *Brassica napus*, Helsingör, 20-23 March 1995, Denmark, L13

Pouvreau A (1984) Production de semences potagères. In : P. Pesson et J. Louveaux (eds.): *Pollinisation et productions végétales.* INRA, Paris. 471-496

Raamsdonk van LWD (1993) Wild and cultivated plants : the paralellism between evolution and domestication. *Evolutionary Trends in Plants* 7(2): 73-84

Rauber R (1977) Evolution von Unkräutern. Ergebnisse der 1. Deutschen Arbeitsbesprechung über Fragen der Unkrautbiologie und -bekämpfung, 1977, Stuttgart, Hohenheim. *Zeitschrift für Pflanzenkrankheiten und Pflanzenschutz* 8: 37-55

Raybould A, Gray A (1993) Genetically modified crops and hybridization with wild relatives: A UK perspective. *Journal of Applied Ecology* 30: 199-219

Raybould A, Gray A (1994) Will hybrids of genetically modified crops invade natural communities ? *Tree* 9 (3): 85-88

Regal P (1992) Scientific principles for ecologically based risk assessment of transgenic plants. Draft prepared for the symposium Potential Ecological and non-target effects of transgenic plant gene products on agriculture, silviculture and natural ecosystems. November 30-December 2, 1992, College Park, Md

Regal PJ (1994) Scientific principles for ecologically based risk assessment of transgenic organisms. *Molecular Ecology* 3: 5-13

Rhodes CA, Pierce DA, Mettler IJ, Mascarenhas D, Detmer JJ (1988) Genetically transformed maize plants from protoplasts. *Science* 240: 204-207

Rich TCG (1991) Crucifers of Great Britain and Ireland. *B.S.B.I. Handbook* No. 6. Botanical Society of the British Isles, London

Rissler J, Mellon M (eds.) (1993) *Perils amidst the promise: ecological risks of transgenic crops in a global market*. Union of Concerned Scientists, Cambridge, M.A

Rösler L (1969) Zur Bestimmung und Beurteilung fon Fatuoiden und Bastardformen zwischen Saat- und Flughafer. *Saat- und Pflanzgut* 10: 731-735

St. Pierre MD, Bayer RJ (1991) The impact of domestication on the genetic variability in the orange carrot, cultivated *Daucus carota* ssp. sativus and the genetic homogeneity of various cultivars. *Theoretical and Applied Genetics* 82: 249-253

Salm van der T, Bosch D, Honee G, Feng L, Munterman L, Bakker P, Steikema W, Visser B (1994) Insect resistance of transgenic plants that express modified Bacillus thuringiensis cryIA(b) and dryIC genes: a resistance management strategy. *Plant Molecular Biology* 26: 51-59

Scheffler J, Parkinson R, Dale P (1993) Frequency and distance of pollen dispersal from transgenic oilseed rape *(Brassica napus)* . *Transg. Res.* 2, 356-364

Schlink S (1994) Ökologie der Keimung und Dormanz von Körnerraps (*Brassica napus* L.) und ihre Bedeutung für eine Überdauerung der Samen im Boden. *Dissertationes Botanicae* Band 222. Cramer, Berlin, Stuttgart, 194p

Schlüter K, Potrykus I (1995) Horizontaler Gentransfer von transgenen Pflanzen zu Mikroorganismen (Bakterien und Pilzen) und seine ökologische Relevanz. Gutachten BATS, Fachstelle für Biosicherheit und Abschätzung von Technikfolgen des Schwerpunktprogrammes Biotechnologie. 1-39

Scholz H (1983) Die Unkraut-Hirse (*Panicum miliaceum* subsp. *ruderale*) neue Tatsachen und Befunde. *Plant Syst. Evol.* 143: 233-244

Scholz H (1991) Die Systematik der *Avena sterilis* und *A. fatua* (Gramineae). Eine kritische Studie. *Willdenowia*, 20, 103-112

Schönberger H, Vries De G (1991) So bauen Sie den Rapsbestand gezielt auf. In: Rapsanbau für Könner - Bestände gezielt aufbauen und schützen. *Top agrar - Das Magazin für moderne Landwirtschaft*. Landwirtschaftsverlag Münster-Hiltrup, 20-26

Schwaintz F (1971) Die Entstehung der Kulturpflanzen als Modell für die Evolution der gesamten Pflanzenwelt. In: G Heberer (eds.). *Die Evolution der Organismen*, II/2, Stuttgart

Shiba H, Takayama S, Iwano M, Shimosato H, Funato M, Nakagawa T, Che F, Suzuki G, Watanabe M, Hinata K, and Isogai A (2001) A Pollen Coat Protein, SP11/SCR, Determines the Pollen Sspecificity in the Self-Incompatibility of *Brassica* Species. *Plant Physiology* 125: 2095-2103

Simmonds NW (ed.) (1976) *Evolution of crop plants*. Longmann, London and New York

Simmonds NW (ed.) (1979) *Principles of crop improvement*. Longman, London and New York

Simmonds NW (ed.) (1988) *Principes d'amélioration génétique des végétaux*. Les Presses Universitaires de Laval, Québec

Small E (1984) Hybridization in the domesticated weed-wild complex. In : WF Grant (eds.): *Plant Biosystematics* Academic Press, Toronto, 195-210

Sukopp U, Sukopp H (1993) Das Modell der Einführung und Einbürgerung nicht einheimischer Arten - Ein Beitrag zur Diskussion über die Freisetzung gentechnisch veränderter Arten. *Gaia* 2(5): 267-288

Sukopp U, Sukopp H (1994) Ecological Long Term Effects of Cultigens Becoming Feral and of Naturalization of Nonnative Species, *Experientia* 49 (3): 210-218

Summermatter K (1996) Systemisch induzierte Resistenz bei Arabidopsis thaliana. Inaugural-Dissertation, Université de Fribourg, 49 pp

Tabashnik BE (1994) Evolution of resistance to *Bacillus thuringiensis*. *Annual Reviev of Entomology* 39: 47-79

Tabashnik BE, Finson N, Johnson MW (1991) Managing resistance to *Bacillus thuringiensis*: lessons from the diamond back moth (Lepidoptera: Plutellidae). *J. Econ. Entomol.* 84: 49-55

Tabashnik BE, Schwartz JM, Finson N, Johnson MW (1992 Inheritance of resistance to Bacillus thuringiensis in diamondback moth (Lepidoptera: Plutellidae). *J. Econ. Entomol.* 85: 1046-1055

Thacker JRM (1994) Transgenic plants. *Tree* 9: 486

Theisen H (1991) Bio- und Gentechnologie - Eine politische Herausforderung. Kohlhammer, Stuttgart, 193

Thellung A (1912) La flore adventice de Montpellier. *Mem. Soc. Nation. Sci. Nat. Math.* 38: 57-728

Tiedje JM, Colwell RK, Grossmann YL, Hodson RE, Lenski RE, Mack RN, Regal PJ (1989) The planned introduction of genetically engineered organisms: Ecological considerations and recommendations. *Ecology* 70 (2), 298-315

Timmons AM, Charters YM, Crawford JW, Burn D, Scott SE, Dubbels SJ, Wilson NJ, Robertson A, O'Brien ET, Squire GR, Wilkinson MJ (1996) Risks from transgenic crops. *Nature* 380: 487

Tomiuk J, Loeschcke V (1993) Conditions for establishment and persistence of populations of transgenic organisms. In: K Wöhrmann, Tomiuk J (eds.): *Transgenic organisms* 117-133

Umbeck PF, Barton K, Nordheim E, McCarthy J, Parrott W, Jenkins J (1991) Degree of pollen dispersal by insects from a field test of genetically engineered cotton. *J. Econ. Entomol.* 84: 1943-1950

Vries de FT, van der Meijden R, Brandenburg WA (eds.) (1992) Botanical files - A study of the real chances for spontaneous gene flow from cultivated plants to the wild flora of the Netherlands. *Gorteria* suppl.1, 100p

Vries de IM (1990) Crossing experiments of lettuce cultivars and species (Lactuca sect. Lactuca, Compositae). *Plant. Syst. Evol.* 171: 232-233

Weber B (1995) Überlegungen zur Aussagekraft von Risikoforschung zur Freisetzung transgener Pflanzen. In: S Albrecht, Beusmann V (eds.): *Ökologie transgener Nutzpflanzen, Gentechnologie - Chancen und Risiken.* Campus.Verlag Frankfurt/New York. 31: 111-126

Welten M, Sutter R (eds.) (1972) *Verbreitungsatlas der Farn- und Blütenpflanzen der Schweiz.* Vol. 1 & 2, Birkhäuser, Basel

Wijnheimer EHM, Brandenburg WA, Ter Borg SJ (1989) Interactions between wild and cultivated carrots (*Daucus carota* L.) in the Netherlands. *Euphytica* 40, 147-154

Williams IH, Corbett SA, Osborne JL (1991) Beekeeping, wild bees and pollination in the European community. *Bee World* 72: 170-180

Williams SL, Dincher S, Carozzi N, Kessman H, Ward E, Ryals J (1992) Chemical regulation of *Bacillus thuringiensis* alpha-endotoxin expression in transgenic plants. *Bio/technollogy* 10: 540-543

Williamson MH (1988) Potential effects of recombinant DNA organisms on ecosystems and their components. *Tree* 3: 32-35

Williamson MH (1993) Invaders, weeds and the risk from genetically manipulated organisms. *Experientia* 49: 219-224

Williamson M, Perrins J, Fitter A (1990) Releasing genetically engineered plants: Present proposals and possible hazards. *Trends in Ecology and Evolution* 5: 417-419

Wrubel RP, Krimsky S, Wetzler RE (1992) Field Testing Transgenic Plants. An analysis of the US Department of Agriculture's environmental assessments. *BioScience* 42(4): 280-288

Young S (1989) Wayward genes play the field. *New Scientist* 49-53

Zohary D (1992) Reproductive biology - introduction. Council of Europe, first workshop on conservation of the wild relatives of European cultivated plants: developing integrated strategies. Ría Formosa Nature Reserve, FARO (Portugal)

[illegible] (1990) [illegible]
[illegible] (1993) [illegible]

© 2003 Birkhäuser Verlag Basel/Switzerland

Genetically engineered crops and sustainable agriculture

Hans R. Herren

International Centre of Insect Physiology and Ecology (ICIPE), P.O. Box 30772, Nairobi, Kenya

Clarification

I have nothing, nor does ICIPE, against the science or principles behind biotechnology, or genetic engineering for that matter. Both have potential to improve agricultural production in the framework of sustainable development, and indeed the former has a proven track record of applications in tissue culture and marker assisted breeding. The question I would like to pose is: How much and when, for what purpose and under what conditions, and at what cost to other agricultural/biological/ecological disciplines can we really expect from these technologies in the short and medium terms?

Background

There has been a tremendous amount of discussion and writing on the biotechnology/genetic engineering issues over the past three to four years. Unfortunately, the bulk of this has consisted of repetition, accusations and polarization with little in terms of real scientific progress. A major reason for this is that on the one hand the biotech/GE industry is, by nature, pressed to provide marketable products to get returns on its investments rather than to concentrate on detailed impact studies, and on the other, the regulatory processes, policy development, impact and deployment research have been left grappling with ever-decreasing overall funding for development cooperation, agricultural research in particular (Pinstrup-Andersen and Cohen, 2000). FAO has taken the lead in opening up the discussion on the positive and negative aspects of biotechnology's potential to feed an increasing world population (http://www.fao.org). It clearly states that "Biotechnology, together with other technologies, could provide new solutions to some of the old problems hindering sustainable development and achievement of food security". I believe that not enough attention is being paid to the words "together with other technologies", a point made very clear by Dr. Ismael Serageldin, in *Ag Biotech and the Poor;* although one hears the echo of that statement everywhere, its actual meaning has been lost.

When one looks at the issue of the negatives of GE, or better put, of the potential risks of GE, there are again only small bits and pieces of information that surface, showing that the research on that end of the spectrum is little, much of it in its infancy (especially in the tropics) and certainly not in tandem with the production (positive side) research. This is unfortunate, as the deployment of GE crops, within the internationally agreed biosafety framework, will depend on the full knowledge of the potential, or real risks involved, if any. But that will only emerge from solid, serious and independent research that will provide the base line data, as well as the manpower for the national decision making processes. Gene flow research is just one example of what needs to be done now. Therefore it is important to, on the one hand, continue investments in new technologies, their biosafety aspects, deployment strategies, regulatory aspects and socio-economic implications, in an overall balanced agricultural and environmental investment strategy.

Going through many papers, in particular the proceedings of the latest gathering in Washington to discuss Biotechnology and the Poor, there is no clearly defined strategy on how the public and the private sectors alike, will proceed to realise the many recommendations on what I call the "other side of the coin" research. There is actually fear in many well respected quarters, that the present trend in international agricultural research (globally, but also in the developing countries) is increasingly being biased towards GE, and will not adequately address the immediate needs of the small farmers or poor consumers (Pinstrup-Andersen and Cohen, 2000). There has been much discussion over the past years on the need to take a holistic approach to sustainable agriculture, given the many drawbacks encountered in 'modern', high-input agriculture. Most of us will acknowledge that this broader view includes research into the deployment of biotechnology products (now) and genetically engineered crops (later, when all the biosafety, deployment and socio-economic issues have been satisfactorily answered).

Are we learning from the past?

The public sector and the consumer are approaching the GE issue with caution, most likely based on the experience we all went through over the past 50 years with pesticides usage. The discovery of DDT and, later, a plethora of new and very effective compounds to kill insects, fungi, nematodes, mites, etc. created the illusion of a wonderful, and (as was then claimed) safe, solution to all pest problems. Promising research in biological control was thwarted, as an analysis of the publications relating to these two fields of pest control shows. Farmers were hooked onto a treadmill of ever-increasing use of more and more potent pesticides, until a member of the civil society raised the alarm: Rachel Carson in the book *Silent Spring*. This best-seller triggered a new era of research and the pioneering work around the world of scientists such as Perry Adkisson, Ray Smith, Paul de Bach, R. van den Bosch, Vittorio Delucchi, Joop van Lenteren and others. Integrated pest management was born, and biological control, which was almost abandoned by the 'quick fix' mentality that came with DDT, again had a new lease of life.

All of the early IPM efforts were funded from public sources, and only later in the day by private industry, when it had no other choice than to join the bandwagon. However, it is well known to IPM practitioners that the industry has modified the IPM paradigm to its advantage, and only in the last 10 years made genuine efforts (under pressure from the public and policy makers) in espousing the ecological approach recommended by the fathers of IPM. Also, under pressure from the public sector, and in particular UN agencies such as FAO and UNEP, important sums of money are now being applied for the removal and disposal of huge amounts of unused pesticides stockpiles in Africa. The industry, the major beneficiary from the sale of these products, has been slow in helping, and is contributing only little to the effort, straining even further the small budgets for development cooperation. Given such a track record, who would blindly follow the lead of the industry (even with its new name 'Life Science Industry')?

We need industry in the quest for solutions to world hunger, poverty and health for all. Industry, however, has as its main objective profit, not philanthropy nor development aid. This is a given, and therefore civil society needs to do its part by supporting governments that promote development cooperation and keep an eye on the industry to assure that it, too, contributes its part to global-wellbeing. There is no doubt that the industry has also contributed to the global food supply, a fact that needs to be acknowledged to its full merit. One can however ask, "Could it have been done in a better way?", and I would say, yes. Can we learn from the past? – I say we absolutely must.

What is the way forward?

The foregoing examples point towards the need for a balance between public and private sector research, and that governments around the world need to re-invest with increased overall levels into agricultural research, never mind the present glut of food in the developed countries. This research will need to cover the broad range of disciplines required for sustainable production, in particular natural resource management that will ensure that high yielding seeds will find an environment that will promote the realization of their potential. A clear strategy for sustainable agricultural production is also still lacking although new attempts at integrated development are being experimented with. Examples would be the BioVillage experiments in India and in Africa, that have started independently, but with one common denominator: they are community-initiated projects that build on know-how from modern science and traditional knowledge.

Alternatives to GE crops: immediate solutions

A novel utilization of biodiversity to control stemborers and an intractable weed in Africa, called the 'push-pull' system, links the utilization of naturally occurring plant and insect characteristics to manage both pests and weeds in maize and sorghum production systems. Wild grasses are used to either attract or repel stemborers from the maize field, while a legume plant is used to repel borers, attract natural enemies, and control the witchweed (striga). The system now being adopted by thousands of farmers in Kenya is being requested by the national agricultural research systems (NARS) of Ethiopia, Uganda, Tanzania, and all the way to South Africa. The costs to the farmers are minimal, as they have always grown the wild grasses, and the legume, as fodder for their livestock. They have the opportunity to use their preferred maize varieties (OPVs and hybrids), need no, or only little, mineral fertilizers (as manure is applied back to the fields), fight erosion and escape minor drought stress through improved soil characteristics and improve their yields up to 30% in the first year and more in subsequent years, as soil fertility builds up. The Gatsby Foundation-funded project has been so successful, that new income-generating activities have taken hold in the area of implementation, women having started seed production of the legume fodder. The project is investigating biotechnological aspects, and possibly one day, GE. Thanks to a Rockefeller Foundation grant, ICIPE is investigating the anti-striga properties of *Desmodium* sp. that could lead to the development of a product which farmers that have no need for fodder may use. Advanced research is going on into understanding the wild grasses' chemicals at the origin of biochemical/behavioral activity on the insects. New projects are being developed by ICIPE with partner national agricultural research and extension systems (NARES) across Africa, on exploration of wild grasses and potential of new discoveries in terms of insect control genes and other useful agronomic characteristics (funded by the Global Environmental Facility). But at least while we are doing this, we have a push-pull solution for the farmers, which is spreading like wild fire. In addition to the push-pull technology, ICIPE has also introduced parasitoids against an exotic stemborer, *Chilo partellus,* and the models predict that it will control the borers at a level that, in conjunction with agronomic practices, will make them economically irrelevant. All this for less than US$ 5 million for some 14 countries of East and southern Africa, and at practically no recurrent costs for the farmers.

When looking at alternatives, let us also remind ourselves of the past successes that one tends to forget instead of using as models. The cassava mealybug is a case in point, in which an entire crop and 200 million people were threatened across the African continent. Millions of farmers had given up growing cassava, and governments had ordered the burning of whatever crops were left. Thanks to the foresight of many donor agencies, in particular the Swiss, German and Austrian government aid agencies, funding was made available for what turned out to be one of the largest (surface covered) and most successful (cost/benefit of 1/240) development and biological control programs so far. Not only were some 20 million people saved from starvation, a crop was saved, an enormous amount of pesticides was not used, and biological control made a resurgence on the African continent. In its wake, 1200 entomologists and technicians were trained, and now form the backbone of plant protection services throughout the continent. Many IPM projects in maize, cowpea, banana and cassava are led by 'graduates' of that program and of the ICIPE postgraduate program in insect science.

Unfortunately the wind has gone out of the IPM/biological control sail, as funding for overall agricultural research and specifically for classical disciplines, shrinks, as many young scientists, including at ICIPE, flock to the biotech labs, where prospects of a career are better than in taxonomy or insect physiology laboratories. At ICIPE we are trying to balance the research activities according to a holistic strategy that is geared toward achieving our mission statement of promoting the well-being of the peoples in the tropics through insect science and capacity building in support of sustainable development. This strategy includes collaboration with public and private sector institutions, both in the developed and developing nations.

Socio-economic/environmental context

In Africa, the vast majority of the farmers are poor, lack access to credit and often good land, and have difficulties accessing inputs and then markets. Farmers also miss the assistance of extension services. NGOs have moved to fill the void in some countries while in others (e.g. Ethiopia), the extension service has been revitalized but lacks access to information and knowledge (decision making tools) and good agricultural education. Africa has, by its geographical, morphological and climatic characteristics, a far more difficult path to follow for its development than other continents. Africa has also one of the biggest human health problems, which compounds the development difficulties. With this background, one can see that in order to move ahead, there is a need to tackle several issues simultaneously, in particular health and nutrition (agricultural production), both of which are linked to poverty alleviation.

What is the best way forward? It is an integrated approach that will cover these issues, such as ICIPE's BioVillage approach, where the development partners get together and assist the communities with good science and solutions that span from production to marketing. Unless money is brought into these communities, there is no way forward, and GE or other high input technologies will not take root. The solution is to take care of these pre-conditions and create an environment that is conducive to new technologies being adopted, because there is an economic benefit. So far little has been done in that direction. Governments, public and private institutions need to work together on these issues, should farmers be expected to invest in inputs (such as improved seeds). We know that the farmers of the green revolution needed to access water, fertilizers and pesticides. These elements, together with the high yielding seeds, were the key to higher production. In the African context, a green revolution has been frustrated by the very diverse environment, lack of appropriate infrastructure, draught animals (mostly afflicted with tsetse and tick-borne diseases) or mechanization, as well as poor government policies. These constraints will need to be addressed before the farmers will invest in high yielding seeds. Who wants to drive a Formula One race car on a road full of potholes? To infer that good seeds would solve the problem of food security is a dangerous oversimplification that promotes one-track thinking and investment. The issue is being acknowledged, but little is done about it. The 'quick fix' mentality has resurfaced and needs to be dealt with in a timely and appropriate manner.

Public vs. private sector research, and access to the common goods

The debate over public vs. private sector research may be relevant in the short term, but given the globalization trend and the move to privatize research around the world (World Bank recommendations to the NARS), there is little doubt that in the long term all research will be in private hands, as will be seed production, and perhaps the most relevant genetic resources. We already know that a large amount of genetic material originates from the developing countries, and all the efforts to conserve germplasm being undertaken by the IARCs, as laudable as they may be, will go only so far. The philanthropic moves by some foundations to support public sector research may be well intentioned and useful, but what can these efforts achieve in the wake of globalization/privatization. I do not have answers on what will happen, but unless there is a major change in attitude by both the governments and the civil society to take charge of what are considered basic needs, we will grow and eat what is being sold to us, not necessarily what we would like to. We all know that access to healthy and nutritious food, just as health, is a basic human right, and so is the right to choose what we want to eat. This has little meaning in the eyes of the average stockholders, again making exception for the few, socially and environmentally conscious ones.

I am an optimist, one that has worked in Africa for close to a quarter of a century. I can say that things are changing for the better, even if at a slow pace. Farmers all around are keen to learn and adopt new technologies. It is our responsibility as agricultural scientists to help them with technologies they can afford and that will assure them and their children a better future.

References

Pinstrup-Andersen P, Cohen MJ (2000) Modern Biotechnology for Food and Agriculture: Risks and Opportunities for the Poor. In: Perseley GJ, Lantin MM (eds.): Risks and Opportunities for the Poor. CIGAR, Washington, 159-169

Methods for Risk Assessment of Transgenic Plants
IV. Biodiversity and Biotechnology
K. Ammann, Y. Jacot and R. Braun (eds)
© 2003 Birkhäuser Verlag Basel/Switzerland

Technology, sustainable agriculture and biodiversity: the socio-economics of technology transfers

Willy De Greef

IBRS, Kortrijksesteenweg 748, 9000 Gent, Belgium

Last year the Subsidiary Body on Scientific, Technical and Technological Advice (SBSTTA) of the Biodiversity Convention (CBD) ranked the priority threats to global biodiversity as follows: first comes habitat loss (most of it through the expansion of cultivated land), second comes the introduction of exotic species. The significance of this SBSTTA statement goes much beyond the actual ranking. It lies in the fact that the CBD has started to act as an environmental management body, by acknowledging that in a world with limited resources, problems have to be solved according to priorities. In other words, it has moved from the first step in problem solving (identification of problems) to the second one (ranking of problems).

Both of the top ranking issues are closely linked to agriculture and food production. It comes as no surprise that the ecological footprint of human agriculture is one of the most important ecological disturbances on earth, as it struggles to cope with feeding a population that has expanded fourfold in numbers in less than a century, and more than that in wealth (and therefore use of raw materials, including energetically more wasteful forms of food).

The purpose of this paper is to visit the challenge of perpetuating their performance while reducing their side effects on the environment, especially biodiversity.

Who will feed the world in 2050?

The challenge for this system in the next half-century is in one aspect less arduous than over the past 50 years: we now dispose of a massive amount of demographic, agronomic, socio-economic and environmental information that was not available to long term planners in 1950. We also have a body of scientific research on the impact of innovation that allows us a more integrated approach to the coming problems. This makes it possible to attend to a number of parameters in our long range planning that our predecessors fifty years ago had to sacrifice by lack of data and for the sake of achieving the primary goal: increasing global food supply by 2-3% per year, in a sustainable way.

We can definitely feed a world with 9 billion people who are much better off than today, with a more varied and nutritious diet. This is not to belittle the enormous effort that will be needed to make it happen in practice, and certainly a lot can go wrong (the unpredictable effects of climate change being one of the major imponderables), but by and large we do have the knowledge and management resources to do the job. The question is: who will do it, and how sustainable will it be?

The answer to that question requires consideration of factors that go far beyond the actual "raw agronomy" of the global food issue. They include considerations on environmental impact, regional food security, social fairness, and agriculture as an agent for income distribution. All this comes over and above the need to increase production globally while at the same time reducing reliance on the impact of a number of unsustainable inputs which are at the moment irreplaceable, irrigation, fertiliser use, crop protection chemicals, mechanical and energy resources.

We should get away from the ideological thinking on these inputs. They have saved the world from mass starvation and the predictable social upheaval that could easily have been our fate if they had not become available to the agricultural community over the past 50 years. Nor is there a case to be made for their complete removal from the system in the future. What is needed is to reduce their ecological footprint, and to identify and eliminate those aspects that are clearly unsustainable over the long term (e.g. the use of "fossil water" resources in irrigation), and those that could compromise future productivity of the agricultural environment (e.g. erosion).

Given the many essential components of change towards a sustainable agriculture in the 21st century, the current debate on agricultural biotechnology and sustainable agriculture seems primitive and certainly uninformed. We have so many key components of our present agricultural systems to change, phase out or replace, that a political decision to eliminate one technology in a "precautionary manner", as is undoubtedly being attempted by some decision makers for biotechnology at the moment, seems irresponsible. Moreover, the recent trend towards reduction of agricultural assistance to those parts of the world where population increase requires the most rapid increase of resources, is rapidly leading to increased dependence on food production in the industrialised world. Between 1986 and 1994, the total amount of agricultural assistance to developing countries slumped from 19 to 10 billion USD (IFPRI). During this period the developing world population increased by 600 million people, that of the developed world by 60 million (FAOSTAT).

Over the next 30 years, it is estimated that up to 90% of the population increase worldwide will be in third world urban centres. Who will feed these cities? These people buy their food, they do not grow it. This observation has profound consequences for future agricultural policy. It means that the present emphasis on assistance of subsistence agriculture in LDCs, while addressing a real need, does not solve the most pressing future problem of these countries. The real solution lies in a commercial agricultural system that is able to feed the emerging urban population from the regional rural areas. For that system to be up to the task, it needs to be productive, low cost and sustainable. If not, these cities will be fed by the low cost, highly productive agriculture of the "farming superpowers": North America, the EU, parts of Latin America.

Is it production or distribution?

For a couple of decades now, the "production vs. distribution" polemic about food security has raged. In other terms: is the achievement of global and individual food security primarily an agricultural challenge or a socio-political one? This is a false debate. It should be obvious that food security at the level of the individual citizen is a distribution problem, and will become more so, as a higher proportion of people will live in cities. However, better distribution of food and income to buy it will only result in higher food prices unless it is accompanied by large increases of food production. At the global level, and over the long term, the simple law of supply and demand will act. The first victims of rising food prices are those who most need it: the resource poor in resource poor countries.

Improving distribution of food supply without increasing production is equivalent to levying a regressive tax on the poor. In highly developed economies like the EU, the cost of agricultural raw materials in the cost of food at the consumer level is estimated to be between 10 and 20%. In LDCs, it is usually closer to 50%. In other words: an increase in the cost of agricultural raw materials has a much higher impact on the poor than on the rich. The reverse is also true.

Table 1: Real farm prices of maize and soybean this century ($/tonne) (source: USDA)

Year	Maize	Soybean
1910-1920	350-400	700-800
1950-1960	225-275	600-700
1990-2000	90-110	200-250

Table 2: Population and cereal supply components 1961-1998 (all data in % of 1961 figures) (source: FAOSTAT)

Country	Area harvested	Yield	Production	Population
Asia	+19%	+153%	+202%	+102%
Africa	+63%	+50%	+146%	+154%
Latin America	+22%	+121%	+172%	+112%

Lowering the cost of agricultural raw material by a more efficient and more productive agriculture has a higher impact on the poor. For example, the farm gate price of maize and soybeans in the USA have been reduced by a factor three between 1910 and 2000 (Tab. 1). This is mostly a consequence of the introduction of more productive agriculture.

Further increases in the physical availability of food can be achieved through only three means. We can cultivate more land, we can increase yield on existing farmland, and we can reduce post-harvest losses. Every single measure to increase global food supply, whether of a technical, policy, economic, social or business nature, can only act upon one or more of these production factors. The biggest environmental disaster that could have struck us over the past century would have been if we had to feed the fourfold population increase through increase of farmland alone. The fact that we did not have to do this, we owe mostly to the deployment of new technology developed since the middle of the 19th century.

An example is the green revolution (Tab. 2). Whatever is said about the side effects of the Green Revolution, it is not seriously disputed that it contributed massively to food security in Asia and Latin America. In contrast, it is widely held that it failed to make a sufficient impact in Africa. The data in Table 2 seem to bear this out. Between 1961 (the first year for which FAO global data are available) and 1998, both Asia and Latin America have achieved cereal production increases massively larger than their population increase, and they have achieved most of that increase through yield increases. Africa has barely been able to keep up with population increase, and most of this was through new land cultivation.

The picture is not much different for root and tuber crops (mainly cassava and sweet potato). For these crops, Africa has been able to keep up with population increase, but at the cost of a massive increase in cultivated land (Tab. 3). Asia and Latin America are essentially phasing out these crops as staples.

Root and tuber crops present a number of disadvantages as commercial commodities. Unless they are processed close to the production area (essentially drying), they are bulky (consisting mainly of water) and therefore more energy-costly. Drying itself is energy-intensive. They also tend to be more subject to post-harvest degradation if stored in suboptimal conditions.

So if we consider the three main components of food physical availability again, it seems as if Asia and Latin America are by and large on track to achieve higher production through yield increase, while Africa does most of it at the cost of habitat, and mainly with crops that have an inherently higher risk of post-harvest loss and a handicap for efficient distribution.

Table 3: Population and root and tuber crop supply components 1961-1998 (all data in % of 1961 figures) (source: FAOSTAT)

Country	Area harvested	Yield	Production	Population
Asia	+0%	+107%	+106%	+102%
Africa	+130%	+38%	+220%	+154%
Latin America	+21%	+18%	+45%	+112%

Increased knowledge inputs in long range agricultural policies, through more informed planning, rapid adoption of technology and better distribution, can and should be able to avoid an extrapolation of these damaging policies far into the future.

The role of science in the future

It can be argued that we do not need additional scientific innovation to keep improving the global food supply while decreasing its ecological footprint. Wider adoption of existing knowledge and policies can achieve more in the short term. That is correct, but planning for agricultural sustainability is only relevant if we consider the long term. The present glut on the agricultural markets has many causes, but there is little doubt that it has for a large part been caused by technological innovation creating productivity increases that consistently outpaced the increase in demand. The result is cheaper food than we ever had for the large majority of the world's population.

The low food prices do bring their own problem: rural poverty for the less competitive farmers. But rural poverty can be addressed through a number of policies. Emptying the R&D pipeline however (as the reduction in public research in developing country agriculture is doing) is guaranteed to bring long-term structural shortages, which will cause both increased poverty, and increased environmental degradation as a stretched agriculture tries to catch up with expanded land use and less sustainable production methods.

Who are we to decide?

At the end of the day, the most important question is: who will take the decisions? Last year in May, the Nuffield Council for Bioethics released a study on agricultural biotechnology and ethics. Its key recommendation regarding biotech in developing countries is: we have a moral duty to make this technology available to the developing world. It would be immoral in a world where so many are still chronically malnourished, to exclude any technology from the range of potential contributors to the solution.

However, we must at the same time make available the means to exercise informed choice about the adoption of technology to developing world. It is not up to us to make choices on their behalf. The Nuffield Council report underpins the need for an effective and inclusive technology transfer system, i.e. not only the technology as such, but the knowledge to bring it into the economy and to supervise its deployment under the same standards of supervision and safety evaluation as in the developed world, without neglecting the different priorities of the recipient countries.

Session 2: Social, ethical and legal issues

Methods for Risk Assessment of Transgenic Plants
IV. Biodiversity and Biotechnology
K. Ammann, Y. Jacot and R. Braun (eds)
© 2003 Birkhäuser Verlag Basel/Switzerland

The cultural implications of biodiversity and biotechnology: an underdeveloped area in the public understanding field?

Jill Turner

The Medical Biology Center, Queen's University of Belfast, 97 Lisburn Road, Belfast, Northern Ireland BT9 7BL

Introduction

Biodiversity, biotechnology and culture are individually contested and intricate concepts. Any discussion which seeks to explore their conjunction risks compounding the complexity. Nevertheless, this nexus is a real and important one as the terms routinely impact upon one another in the lifeworld beyond the academy and policy-making environments. I argue that, thus far, the relationship of culture to biodiversity and biotechnology has been largely neglected in the orthodox field of public understanding of biotechnology.

This paper, therefore, concentrates on notions of culture as these relate to biotechnology and biodiversity. However, as Raymond Williams noted, culture is always open to a multitude of readings. One possible avenue of investigation is to view 'the cultural' in an anthropological sense. This often takes the form of a case study approach. Such attempts are normally presented from particular, albeit unreflected, ideological standpoints of actors in a specific local or regional context.

For example, a farmer, villager or scientist, and though very valid, does not form the content here. Instead, I attempt a meta-analysis which considers culture as popular culture shaped by the culture industries.

Several points are considered. Firstly, the way that culture and biodiversity and biotechnology mutually constitute each other. The world interacts with texts and the social world with cultural representations. Ideology both reflects and produces cultural representations of biodiversity and biotechnology. Still, our recognition of the importance of culture is poorly developed and little explored in the many debates around lay and professional understandings of biodiversity and biotechnology issues. The rich insights gained from cultural studies remain largely untapped by orthodox interest groups concerned with the public understanding of science, biodiversity and biotechnology.

Instead, current discourses tend to focus on scientific concerns or on bioethics. While the social is included in bioethical considerations it tends to be framed within a now well established, and though important, repetitive framework which concentrates upon rights and responsibilities. Although one might expect this of moral philosophy, such routine emphasis results in the marginalisation of social and cultural issues and concerns to the detriment of a more comprehensive appreciation of the rich context within which the public understanding of biodiversity and biotechnology develops. This, in turn, is reflected in the superficial and frail public understanding strategies which fail to grasp many deeper cultural processes affecting public perceptions.

Bioethics has become an established and important area as new or reordered moral questions arise from the emerging technologies that make up the fields of biodiversity and biotechnology. Firstly, bioethics addresses many important moral questions. Secondly, the findings can function as moral guidelines as new frontiers of knowledge, practice and traditions are breached.

The results of bioethics reports are, therefore, potentially powerful legitimating texts. They are purportedly neutral, distanced from the power of science and industry.

Two examples from the bioethics field (chosen for their relevance to this conference and because they are both relatively recent and typical texts rather than because they are particularly remarkable examples) are the Nuffield Council on Bioethics', "The Nuffield Modified Crops Report" and the more critical Novel Food Ethics Council report, "Novel Foods: Beyond Nuffield". Interestingly, both these reports reach fundamentally different conclusions about the use of GM crops. In the words of the Food Ethics Council this took the form of a "yes, but" approach from Nuffield and a "no, unless" response from themselves. Despite their different conclusions both reports assume the use of a relatively ideologically free examination of scientific, legal and social factors. However, an approach using science studies or cultural studies would question that premise. Within these disciplines, largely under the influence of postmodernist and poststructuralist theories, no zones (including their own) are held to be free from ideology. The trick, whenever possible, is to recognise and declare one's own ontological, epistemological and professional perspectives.

The Nuffield Report states that "The report is grounded in liberal, scientific values and takes a broadly utilitarian approach to ethics, *a starting point which is shared by most people in the UK*" (my italicisation).

Now while Nuffield do declare their value system they also unreflectively assume an ideological position which assumes that their own view is the common one. The quote illustrates how the highly ideologically implicated concepts of a dominant value system are "normalised". This simultaneously, if unreflectively, attempts to legitimise science through the incorporation of bioethics as a moral, common value system. This fits well with Antonio Gramsci's concept of "hegemony" which he describes as legitimisation via the realm of ideas and mental production in texts produced by the dominant classes. Here, coercion gives way to consent, "simply the way things are", "the natural order of things". But Gramsci recognised that via culture the possibility of alternative, subversive readings is always present. Many sections of the public in the UK and Northern Europe have become increasingly disaffected with science, biotechnology and biodiversity, often rejecting proffered institutional ideologies. This institutionally acknowledged public disaffection has traditionally resulted in a response which calls for better lay education and an increase in the number of scientists who are able and willing to explain "the facts". The erroneous assumption (thus far) has been that this will result in an increased positive appreciation of biotechnology. This can be summed up as the "to know science is to love it" approach (Turney 1996) but it ignores both the insights noted above and the lay understanding that the same sources and ideology are being repackaged by the same interest groups. To move forward, institutional interests need to reflect upon the sophistication of different lay public understandings, and the role of culture, the culture industries and the mass media.

The role of popular culture and the culture industries

The role of culture is complex and underdetermined. It may be used by both dominant groups and by public interest groups. Audience reaction to, and the reading of, texts in relation to effects has been the subject of much academic debate. It remains uncertain how effective powerful media production and ownership of the culture industries is in achieving desired reactions to particular messages; how much audiences can and do subvert any intended dominant meaning is also open to question despite Gramsci's claims.

Both Adorno and Horkheimer from the Frankfurt School were pessimistic about mass culture's critical role. Adorno and Horkheimer's "Dialectic of Enlightenment" first published in

1947, was a critique of enlightenment rationalism which they saw as being complicit with totalitarianism and the "administered" societies of late capitalism. This work included the joint study, "The Culture Industry: Enlightenment and Mass Deception" where they described the "conformist banalities of a manipulative mass culture".

Adorno looked at the form of achieving autonomy in modern literature and music – rather than the traditional agency of the working class of classical Marxist theory. In "The Culture Industry", Adorno saw the erosion of political consciousness and the absorption of all but the most "authentic art".

Later Marcuse put a more optimistic spin on this in "Eros and Civilisation" and "One Dimensional Man". Via Marxism and psychoanalytic theory he sought a *rapprochement*. His work became key texts for the emerging counterculture of the 1960s. Students and countercultural movements became part of the new political refiguration and in 1978 his work "The Aesthetic Dimension" attributed a new critical role to art (Brooker 1999).

The table of binary oppositions below maps two contrasting "stories" of biotechnology which vie for dominance in public readings (Tab.1).

Table 1.

Pro-Biotech	Anti-Biotech
Predictability	Unpredictability
Control of nature	Loss of control
Natural (enhanced)	Unnatural
Utopian	Dystopian
Private profit	Universal need
Promise	Threat
Safer, more nutritious food	Risk to human health/environment
Raising of agricultural productivity	Reduction of traditional resources
Food security for the developing world	Increased insecurity of markets, access to seeds and ways of life

These contrasting accounts of the same phenomenon stream out into the public realm of ideas. Cultural critics and scholars have offered disparate theories on what happens after that. For Lyotard, the previous authority of science as a grand narrative is over, now plural flowers bloom. Truth has become a relative concept.

Baudrillard describes an implosion of culture where more and more information has less meaning. Jameson argues that mass culture has resulted in an erosion of political consciousness with a consequent lowering of the moral ground tone.

Whatever else can be discerned from discussions within the academy it seems that the "structures of feelings" that Williams describes is at the least sceptical towards science and biotechnology. Culture and the culture industries, are it seems, worthy of contemplation as key factors whether or not academic theorists take a pessimistic or optimistic view of their role.

New technologies such as the internet are changing the cultural landscape and offer transnational and globalised communication networks. A look at some of the sites of the non-governmental organisations are a testament to the ways in which undifferentiated publics can effectively come together to resist particular institutional agendas. They have sophisticated websites offering links to other organisations with a similar agenda. They invite direct action to influence science, food and agricultural policy nationally and globally. The public are able to access a plethora of information in ever unprecedented ways. Any isolated or semi-isolated "messages" coming from those concerned to educate the public about science face a different world from that of the 1980s when science communication and public understanding first came to the fore after cutbacks in science funding.

Popular culture and the arts

The arts and popular culture form a space where science and biotechnology are mediated with public response. The visualisation, myths and allegorisation of biotechnology continue to be produced within popular culture and the arts. Within film studies we can track early films such as "Frankenstein" and "Dracula" through to more recent films such as "Alien"/ "Aliens": with cross-species incubation and reconstruction from DNA; "Gattaca" with a mind/body fascism focussed upon genetic screening; "The Matrix", where human beings are grown not born; "Species", with the cross-breeding of transgenic rat/human. Films such as "The Fly" and "Jurassic Park" point to a lack of regulation, to financial profit, to a lack of precaution, to an inability to predict consequences, and to a threat to the social and moral order. The narratives may, within the constraints of the science fiction genre, be different but the markedly consistent thematic motifs point structurally to an overarching concern: the threat of science.

A glance at dystopian fiction offers a similar representation. Apart from the infamous "Brave New World" and "Boys from Brazil", we can see from Dick's "Do Androids Dream of Electric Sheep?", set in a post-nuclear environment where animals are all but extinct, Marshall Smith's "Spares" where the rich farm clones for body parts and Noon's "Pollen", where transgenic variations of human-canines meet a hay fever epidemic, that science fiction remains committed to narrating the possible darker contours of our culture. Within many postmodern fiction texts the writers are more interested in the social and institutional consequences of scientific endeavour rather than in the science and technology itself. They offer bleak motifs of a world which has been engulfed by international conglomerates and is on the brink of breakdown.

Turning to photography and the fine arts it seems that galleries frequently have exhibits in their collections which tell stories of the dystopian nature of science.

In 1996 Aziz and Cucher held an exhibition at The Photographer's Gallery in London named "Unnatural Selection". Using digital photography they reworked the human image to create a fictional society where technological influence had dehumanised the body. The created citizens of Dystopia had lost their means to communicate or to express emotion because they were trapped in their own hermetically sealed bodies, powerless and impotent (Ruth Charity from programme notes).

In March 2000 at the Glasgow Gallery of Modern Art two representations of science were in evidence. The first was of Love Canal – a photograph with an accompanying text. The text outlined the Dioxin disaster and the negative role the government and the US army played. The representation was more positive – a video and stills from the Royal Society of Chemistry in conjunction with Glasgow Council entitled "109 – Table of Elements", an audio visual exploration via images and animation. This was a highly effective, visually stunning piece of work accompanied by a complementary soundtrack; the result was an aesthetically enjoyable and informative approach to science education.

Discussion

The Glasgow exhibition was notable because of the contrasting representations of science, two different aspects of science competing for public attention amongst an array of other unconnected works. But it was instructive in so far as two stories were being told within the same space.

The above snapshot, to use a photographic metaphor, indicates the enduring appeal and fascination of science and biotechnology to the fine arts and popular culture. It appears that the role of art can be an influential, even a critical one. Turney (1998) assumes such a critical role for art largely through the mythic role of key texts such as Mary Shelly's "Frankenstein".

He asserts the need for new stories and suggests that though Frankenstein has been an enduring and powerful myth it has had its day. Still, Branagh's "Frankenstein" remains readily available on the video shop shelves alongside Alien Resurrection, and I would argue that the myth persists and permeates our culture.

The more interesting questions are "Who will create alternative new stories?" and "Are reassuring stories the most interesting or the most believable ones?" The continued success of horror, science fiction and disaster genres suggest a public predilection for dark tales.

Following the above discussion it may be instructive to think about the impact of popular culture and the culture industries on public understandings when compared to current and earlier efforts made by those in the fields of public understanding of science.

For example, consensus conferences have arguably had a limited impact on large populations. As noted earlier, the World Wide Web offers increasing opportunities and is already effectively used by public interest groups for lobbying and information purposes. The potential of reaching large audiences is writ large in the plans to use on-line voting in Arizona. Innovations here might prove fruitful for public understanding.

Wellcome have been quick to use drama as a means of reaching audiences, first through live theatre and later with video; for example, "The Gift" on the genetics of Freidreich's ataxia and a more recent drama on the subject of cloning.

Christians in the US are beginning to make successful films about Christianity and the experience of being Christian after a long period of disappointments with Hollywood. For example, the recent film "The Christians" offers a model for emulation in other fields which have comparably seen the demise of a "grand narrative".

Conclusion

The biodiversity debate cannot be ring-fenced from the biotechnology/science/risk discourse. An anti-science discourse permeates our culture and "structures of feeling". Calls to inform, to better educate the public, proliferate in pro-biotechnology texts but they largely ignore the importance of culture and insights from Cultural Studies and other Social Sciences. Meanwhile, the refusal to acknowledge more complex analyses of public understanding and audience reaction is accompanied by a growing public scepticism and disdain which proponents of biotechnology ignore at their peril.

A deeper understanding of the cultural context and the culture industries may not translate into an instrumental guide as to "how to win public support", nor should it. However, it should at least take the debate along possible other paths and out of the repetitive, but ultimately unsatisfactory, established patterns of discourse currently operative between alienated publics and frustrated scientific and commercial concerns.

Ignoring such deep cultural anxieties will not make them go away. Recognising them, mapping them and addressing them appears to be a sensible way forward.

References

Adorno TW, in Bernstein JM (ed.) (1991) *The Culture Industry: Selected Essays on Mass Culture.* Routledge, London
Adorno TW, Horkheimer M (1979) *The Dialectic of Enlightenment.* Verso, London
Aziz A and Cucher S (1996) How to Look – Aziz and Cucher See the Future. Great 8, *The Magazine of The Photographers Gallery*, London. Sept/October 1996, London
Baudrillard J, in Poster M (ed.) (1988) *Selected Writings.* Polity Press, Cambridge
Beck U, Giddens A, Lash S (1994) *Reflexive Modernization: Politics, Tradition and Aesthetics in the Modern Social Order.* Polity Press, Cambridge
Brooker P (1999) *A Concise Glossary of Cultural Theory.* Arnold. London, Sydney, Auckland
Cooke L, Wollen P (eds.) (1995) *Visual Display: Culture Beyond Appearances.* Discussions in Contemporary Culture, No. 10. New Press. New York
Crichton M (1990) *Jurassic Park.* Ballantine Books. New York
Dick PK (1997) *Do Androids Dream of Electric Sheep?* Harper Collins. New York
Featherstone M (1991) *Consumer Culture and Postmodernism.* Sage, London
Food Ethics Council (1999) *Novel Foods: Beyond Nuffield.* Food Ethics Council, Nottingham
Gray A, McGuigan J (eds.) (1997) *Studying Culture.* Arnold, London
Gramsci A, in: Hoare Q, Smith GN (eds.) (1971) *Selections from the Prison Notebooks.* Lawrence and Wishart, London
Huxley A (1999) *Brave New World.* Penguin Longman. London
Jameson F (1991) *Postmodernism, or, The Cultural Logic of Late Capitalism.* Duke University Press, Durham, NC
Knorr-Cetina KD (1981). *The Manufacture of Knowledge.* Pergamon Press, New York, Oxford
Levin I (1991) *Boys From Brazil.* Bantam, USA
Lyotard JF (1984) *The Postmodern condition: A Report on Knowledge.* Bennington TG, Massumi B (Transl.), Manchester University Press, Manchester
Marshall Smith M (1998) *Spares.* Harper Collins. New York
McHale B (1987) *Postmodernist Fiction.* Methuen. New York and London
Morley D (1995) Theories of Consumption in Media Studies. In: Miller D (ed.): *Acknowledging Consumption: A Review of New Studies.* Routledge, London
Noon J (1995) *Pollen.* Ringpull. Manchester
Nuffield Council on Bioethics (1999) *The Nuffield Modified Crops Report: Ethical and Social Issues.* Nuffield Council, London
Turney J (1996) *To Know Science Is To Love It?* COPUS by the Royal Society of Science, London
Turney J (1998) *Frankenstein's Footsteps: Science, Genetics and Popular Culture.* Yale University Press. New Haven and London
Wellcome Trust (1998) *Public Perspectives on Human Cloning.* Wellcome Trust Publishing Department. London
Williams R (1976) *Keywords: a Vocabulary of Culture and Society.* Fontana. London

Filmography

Alien, dir. Ridley Scott (1979)
Aliens, dir. James Cameron (1986)
The Fly, dir. David Cronenberg (1986)
Gattaca, dir. Andrew Niccol (1997)
The Matrix, dir. The Wachowski Brothers (1998)
Outbreak, dir. Wolfgang Petersen (1995)
Species, dir. Roger Donaldson (1995)

Methods for Risk Assessment of Transgenic Plants
IV. Biodiversity and Biotechnology
K. Ammann, Y. Jacot and R. Braun (eds)
© 2003 Birkhäuser Verlag Basel/Switzerland

The role of communication in ecological assessments of genetically modified crops

Thomas E. Nickson

Ecological Technology Center, Monsanto Company, 800 N. Lindbergh Blvd., St. Louis, MO 63141, USA

Introduction

Scientists develop terms, phrases and vocabulary to communicate complex ideas to their peers. Each scientific discipline has some unique aspects to its vocabulary, which to the uninitiated may be difficult to understand at first. For individuals grounded in scientific first principles (the basics of science education, methods and training), comprehension of factual information and facile exchange of ideas occurs after becoming versant in the language. Ideally, terms and phrases are developed and used with the intent to expand knowledge in a scientific discipline.

Genetic modification of plants is a rapidly emerging, multi-disciplinary technology. The first products are commercially available in several world areas, principally North America where there has been widespread adoption by growers occurring at an extremely rapid rate (James, 1998). The product development process (bring genetically modified crops to the market) required, in part, the cooperative work of scientists with skills in numerous scientific disciplines including plant and molecular biology, genetics, biochemistry, toxicology and ecology. In addition to these basic sciences, methods and models of ecological risk assessment have been adapted from the chemical pesticide models for genetically modified (GM) crops (Kjellsson, 1997; OECD, 1993). Scientific progress toward developing these products has required the assimilation of the terms and phrases of individual disciplines, and resulted in the development of some new vocabulary. Terms such as "outcrossing", "genetic pollution" and "secondary effects" are often used as a pejorative to describe GM crops; and they are being widely used in the scientific literature. Furthermore, terminologies familiar to the discipline of risk assessment (e.g. "risk" and "hazard") are being used imprecisely in some scientific communications related to GM crops. For example, it is common to read or hear scientists state, "outcrossing is one of the significant risks posed by GM crops". An explanation for why this is imprecise will be given later.

The purpose of this paper is three-fold. Firstly, it is my intention to articulate an important distinction between building scientific knowledge through dialogue versus promoting scientific opinion through debate. Conscious awareness of this difference could highlight potential factors that are negatively affecting the science-base advancement of the ecological risk assessment related to GM crops. Secondly and primarily, I wish to build on a recent publication in the EC (Van den Eede and Housen, no date), and revisit the use of the key terms derived from ecological risk assessment literature (OECD, 1989; Suter, 1993) as they apply to GM crops. Finally, an attempt is made to encourage scientists involved in ecological risk assessment of GM crops to use language more precisely.

Discussion

Despite early efforts to build science-based guidelines to evaluate the ecological[1] safety of GM crops (NRC, 1987; OECD, 1992), many recent discussions surrounding GM crops have been better characterized as a debate and rather than a dialogue. In light of the controversy surrounding scientific ecological safety assessments of GM crops, it is interesting to compare the definition of "dialogue" with "debate":

Debate: A contention by words or arguments: as a) the formal discussion of a motion before a deliberative body according to the rules of parliamentary procedure; b) a regulated discussion of a proposition between two matched sides (derived from Latin word meaning: to beat at battle) (Webster's New Collegiate Dictionary, 1981).

Dialogue: A conversation between two or more persons; an exchange of ideas and opinions (derived from Greek meaning to converse) (Webster's New Collegiate Dictionary, 1981).

The ultimate objectives of a debate and a dialogue are fundamentally different. Based on these definitions, one can infer that information is freely exchanged in a dialogue where two or more individuals share ideas and opinions in a cooperative manner; whereas a debate is better characterized as a presentation and defense of a set of ideas, opinions or beliefs. One might expect to find debates charged more with negative emotion. In a debate, discussions might occur where the parties engaged are likely to be more intent on promoting their view ("winning") as opposed to achieving a mutual goal such as the expansion of knowledge and understanding. In contrast with a debate, a dialogue is characterized by objectivity and cooperation. Furthermore, terms and phrases may be used less carefully in a debate than in a dialogue. It is expected that the participants in a dialogue would develop consensus definitions of the key terms, phrases and vocabulary required to advance the discussion and avoid confusion. In scientific dialogues, there is a need to frequently revisit definitions of terms as well as data to ensure that they communicate clearly and serve the needs of all the involved parties.

The introduction of GM crops has invoked both dialogue and debate in the public and scientific communities as to the ecological safety and benefits of these products. Early efforts to develop consensus concerning the scientific assessment of ecological safety involved scientists from numerous disciplines (NRC, 1987; Tiedje et al., 1989; OECD, 1992). The conclusions and recommendations of these first reports were subsequently challenged and additional recommendations have been offered (Kjellsson et al., 1997; Rissler and Mellon, 1996). Much of the guidance provided to date for assessing the ecological safety of GM crops is based on chemical, biological and ecological first principles. Nevertheless, many scientists use convenient phrases to describe complex phenomena (e.g. outcrossing) or are imprecise in their use of the terms and phrases derived from risk assessment sciences. This may be a result of the complexity of the problem as well as a lack of awareness of the field of ecological risk assessment as applied to GM crops on the part of some scientists. Kjellsson (1997) provided a condensed summary of principles and procedures for ecological risk assessment for GM crops. This publication listed a useful and general framework for risk assessment as well as an extensive list of experimental methods that should be considered depending on the product. A recent publication in the European Community (Van den Eede and Housen, no date) proposed clearer definitions for the key terms derived from ecological risk assessment literature (OECD, 1989; Sutter, 1993) as they apply to GM organisms.

In their publication, Van den Eede and Housen carefully define the key terms:

[1]A distinction between environmental safety and ecological is made here. This publication is focused on ecological safety rather than environmental which includes safety toward humans. Suter (1993) also makes this distinction. It is felt that human safety issues are better addressed specifically and by those who are more familiar with the issues.

- *Harm*, "is the realization of hazard";
- *Hazard*, "is a property of a substance, an act, phenomenon or process that could cause harm";
- *Risk*, "is a measure of the consequences (economic, health, environment) of hazard in terms of both hazards' *likelihood* and *frequency* and the *magnitude* of its consequences" and "the product of (1) the magnitude of the harm (generated by hazard) and (2) the frequency with which the hazard occurs";
- *Risk Assessment*, "a scientifically based process consisting of the identification and characterization of hazards, the estimation of exposure and calculation of the resulting risk."

These definitions are consistent with those that can be found in Suter's text (1993), and were developed based on the scientifically established principles of ecological risk assessment for chemicals:

- *Analysis:* "a formal, usually quantitative, determination of the effects of an action."
- *Assessment:* "the combination of analysis with policy-related activities such as identification of issues and comparison of risks and benefits."
- *Ecological risk analysis:* "determination of the probability and magnitude of adverse effects of environmental hazards (chemical, physical, or biological agents occurring in or mediated by the ambient environment) on nonhuman biota."
- *Ecological risk assessment:* "the process of defining and quantifying risks to nonhuman biota and determining the acceptability of those risks."
- *Exposure:* "the process by which the temporally and spatially distributed concentrations of a chemical in the environment are converted to a dose."
- *Hazard:* "a state that may result in an undesired event, the cause of risk."
- *Risk:* "the probability of a prescribed undesired effect."

A seeming inconsistency is found in their discussion of hazard characterization and identification. Van den Eede and Housen present a list of environmental hazards associated with GM crops that they claim have been identified. In addition to harmful effects such as enhanced weediness (crop and compatible relatives) and adverse impacts to nontarget organisms that have been identified earlier (Tiedje et al., 1989; Rissler and Mellon, 1996), their list includes:

- "The gene pool has the potential to become altered."
- "The structure or function of the ecosystem has the potential to become altered."
- "Establishment of genes in wild populations (vertical gene transfer)."
- "Gene transfer to microbes (horizontal gene transfer)."

These particular "hazards" are inconsistent with the definition of hazard Van den Eede and Housen state, because they are not *a property* of a substance, *a property* of an act, *a property* of phenomenon or *a property* of process. More accurately, since they are actual phenomena and processes themselves, they cannot be ascribed as harmful in and of themselves. A science-based ecological hazard assessment would clearly identify the property of the substance that is responsible for the harm. Confusion and misunderstanding are created when scientists communicate that GM crops possessing transgenes are harmful by nature and biologically and ecologically different than their traditional counterparts. For example, the phrase "outcrossing from GM crops" is often used to as a reference to some predetermined uniqueness of GM crops. However, scientists who have studied the phenomenon of vertical gene flow from transgenic plants have concluded the current GM crops are equivalent to traditionally bred crops in their potential to outcross (Hancock et al., 1996; Jørgensen et al., 1996). Furthermore, any trait existing in a traditional crop has a similar potential to affect biological and ecological processes (NRC, 1987; Tiedje et al., 1989). As such, biological and ecological processes associated with GM crops are not hazardous by nature, but must be evaluated for their potential to result in ecological harm.

Another apparent incongruity in the use of terms commonly found in the scientific literature associated with ecological risk assessment of GM crops is the use of the term "risk" as a synonym for "chance", "likelihood" and "possibility". As noted earlier, references can be found in peer-reviewed articles to biological phenomena such as gene flow as "risks" e.g. "outcrossing is one of the principle risks from GM crops". In such cases, the implication is that the researcher has come to the conclusion that gene flow is a harm that occurs with some frequency. However, careful examination of many of these papers reveals that no hazard assessment has been conducted according to the classical risk assessment model:

Risk = Hazard x Occurrence (or exposure)

As such, no harmful property has been revealed through experimentation that can be attributed to the biological phenomena. In some cases, a theoretical hazard is described along with a conclusion that a risk has been characterized. More accurate communication in these instances might be to use terms such as "chance" or "likelihood"; or to qualify the risk as being "potential".

In his book, Suter makes an important distinction between assessment and analysis that is relevant to GM crops. This differentiation is significant because it distinguishes between a purely science-based *analysis* and an *assessment* that encompasses scientific data and policy-related knowledge. Applying Suter's definition of ecological risk assessment to GM plants requires inclusion of risk and benefit analysis as well as explicit consideration of agriculture, environmental and trade policies. On this aspect of ecological risk assessment of GM crops, there appears to be consensus around the world.

A carefully conducted scientific ecological risk assessment of GM crops is an iterative process with several components including problem formulation, risk analysis, risk characterization and risk management (Kjellsson et al., 1997). Hazard characterization is an integral component of an ecological risk assessment of GM crops. The potential for harm to result from the release of both the modified plant and the introduced trait must be analyzed using the best available scientific information and methodologies. Some hazards can be identified and characterized using the scientific principles of risk assessment developed for chemicals (e.g. nontarget organism testing). Ultimately, hazards are evaluated in the context of their probability to occur and the magnitude of their effect in both time and space. Risk is then the result of this analysis, which is ultimately balanced with potential benefits to complete the ecological assessment process. In circumstances where the risk is acceptable provided risk management is available, post-registration monitoring would be required (Nickson and Head, 1999).

A final point concerns the standards and expectations of ecological safety of products used in agriculture especially GM crops. Käppeli and Auberson (1999) point out that "genetic engineering can not be made safer than biology itself". Rigorous standards of safety must be applied to all products including those derived through modern biotechnology. The basis for determining acceptable ecological risk is the science-based risk assessment. Furthermore, the ecological risks associated with GM crops are determined based on the biology and ecology of the modified plant and the introduced trait using the traditional systems of crop production as the basis for comparison (Tiedje et al., 1989). As such, communication of risks and hazards associated with GM crops should clearly identify the comparative basis upon which the risks and hazards were evaluated.

In summary, the work of Kjellsson (1997), Van den Eede and Housen (no date), and Käppeli and Auberson (1999) provide an excellent platform upon which to establish a dialogue which could advance the science of ecological risk assessment for GM crops. Applying the best available scientific knowledge of biology and ecology within an established framework of risk assessment has provided a valid basis for ecological risk analysis. These efforts can be enhance by the precise use of terms such as "risk" in scientific publications describing GM crops. Awareness and care on the part of scientists to using terms correctly and precisely defining phrases is an important step to achieving this goal. As such, scientists working in this field need to become aware and practice more precise communication of experimental results to further the dialogue concerning GM crops.

References

Hancock JF, Grumet R, Hokanson S C (1996) The Opportunity for Escape of Engineered Genes from Transgenic Crops, *HortScience*, *31*, 1080-1085

James C (1998) Global review of commercialized transgenic crops: 1998. *ISAAA Briefs* No. 8. ISAAA: Ithaca, NY.

Jørgensen R-B, Hauser T, MikkelsenT, Østergård H (1996) Transfer of engineered genes from crop to wild plants, *Trends in Plant Sciences 1*, 356-358

Käppeli O, Auberson L (1998) How safe is safe enough in plant genetic engineering? *Trends in Plant Science* 3, 276-281

Kjellsson G, Simonsen V, Ammann K (eds.) (1997) *Methods for Risk Assessment of Transgenic Plants, II. Pollination, Gene-Transfer and Population Impacts*. 221-236

Nickson TE, Head GP (1999) Environmental monitoring of Genetically modified crops. *Journal of Environmental Monitoring* 1 (6) 101N-105N

NRC (ed.) (1987) *Introduction of recombinamt DNA-engineered organisms into the environment: Key issues.* National Research Council, National Academy Press, Washington, D.C.

OECD (1989) Report of the OECD Workshop on Ecological Effects Assessment. Organization for Economic Cooperation and Development, *OECD Environment Monographs*, No. 26 Paris

OECD (ed.) (1992) *Safety considerations for biotechnology*. OECD, Paris, 50 pp.

Rissler J, Mellon M (eds.) (1996) *The Ecological Risks of Engineered Crops*. MIT Press

Sutter G W (ed.) (1993) *Ecological Risk Assessment*. Lewis Publishers, Chelsea, Michigan, USA

Tiedje JM, Colwell RK, Grossman YL, Hodson RE, Lenski RE, Mack RN, Regal PJ (1989) The Planned Introduction of Genetically Engineered Organisms: Ecological Considerations and Recommendations, *Ecology*, *70*, 298-315

Van Den Ede G, Housen M (eds.) (no date), *Quantitative Environmental Risk Assessment for Genetically Modified Organisms*, on behalf of the Forward Studies Unit of the European Commission. European Commission. Directorate General - Joint Research Center. Institute for Health and Consumer Protection Food Products Unit, Biotechnology and Environment Sector

Webster's New Collegiate Dictionary (1981) G. & C. Merriam Company, Springfield, Massachusetts

Methods for Risk Assessment of Transgenic Plants
IV. Biodiversity and Biotechnology
K. Ammann, Y. Jacot and R. Braun (eds)
© 2003 Birkhäuser Verlag Basel/Switzerland

Intellectual property rights and traditional knowledge: proposals for action

Susette Biber-Klemm

University of Basel, Faculty of Law, Maiengasse 51, CH- 4056 Basel, Switzerland

Introduction

The two notions of intellectual property and traditional knowledge stand for different worlds, different outlooks on life, and – of importance within our context – for different methods of innovation.

Recent developments in science, industry, and business that have been characterized by progress in biotechnologies and by globalization of markets, have brought the two worlds into contact with each other and enhanced the interdependence between industrial and traditional innovation processes.

The factual and legal questions that arise at the interface of traditional knowledge and intellectual property will be developed in this paper, while analysing problems and submitting possible solutions. Before entering into analysis of the problems, some factual aspects are first described. The paper will then discuss the present legal order and existing instruments at the international level. Based on two case studies, problems and questions arising in implementing these instruments will then be elaborated. In a next step, a draft national law on traditional knowledge, access, and benefit sharing will be analysed regarding its capacity to resolve the problems described.

Aspects of the factual background

Scientific progress in recent decades has been characterized by two important traits: first, by the insight into the interconnection of all ecological processes at the local, regional, and global level, and, related to this ecosystem approach, by the insight into the importance of diversity among animals and plants to maintain ecosystem functions and ultimately the survival of mankind. Second, by the rapid evolution of the genetic sciences and technologies that brought about changes in the perception of life and the living environment and an enhanced understanding of diversity's importance.

Parallel to this process, progress in biotechnology has led to new insights into the use of biological resources. Industry is investing large sums to create new products based on natural resources in the fields of food, pharmaceuticals and lifestyle. This trend is supported by consumer interest. In Western countries this is changing in the direction of environment sensitivity and thus enhances interest in natural products. In the field of food crops, progress in genetics has generated a new understanding of the importance of maintaining a broad genetic basis to assure food security in a changing world (Biber-Klemm, 2000c).

In these fields, the knowledge of farming communities and indigenous peoples plays a twofold role: it is of importance in conserving and maintaining diversity in wild or domesticated species. On the other hand, this kind of knowledge plays an important role in identifying biological resources worthy of commercial exploitation and contributes valuable information to formal breeding processes. Locally adapted landraces resulting from continuous breeding efforts of generations of farming and indigenous communities are an important source of genetic diversity in plant genetic resources for food and agriculture (PGRFA). Thus they are of importance in maintaining food security.

However, animal and plant species are being extinguished at an accelerated rate. This process is influenced by all kinds of human activity and productivity (see Biber-Klemm, 2000a-c). Closely linked to loss of biodiversity is the decline of traditional knowledge.

According to economic theories, an important reason for the loss of genetic diversity and related traditional knowledge lies in its specific informational character: genetic resources are self-propagating and also produced and traded as goods for consumption. Knowledge, once it has been revealed, is accessible for everybody. Thus information contained in landraces and traditional knowledge – from the legal point of view – is in the public domain, i.e. subject to open access. Once the information has been acquired, it is impossible for its author to prove that it was his originally. Because the main value of diversity and traditional knowledge cannot be appropriated, it is therefore ignored by land-owner decisions, and diversity of plant species is sacrificed in favour of more lucrative use of land resources such as industrialised monocultures (Swanson, 1995).

To protect results of the formal, industrial research and development (R&D) processes, specific instruments to protect informational values of inventions or new plant varieties have been created – intellectual property rights (IPRs). However, results of informal and traditional R&D processes characterized by their continuous evolution, such as landraces and traditional knowledge, presently cannot be protected by IPRs, as they do not correspond to the protective criteria.[1]

Evolution in the world of formal innovative processes during recent decades has been characterized by shifting R&D from the public to the private sector. This has had important consequences in regulating intellectual property rights, in particular in domains where creation of new information is an important part of industrial production, as is the case in pharmaceuticals and agriculture (Swanson and Göschl, 2000). In these industrial sectors, high investments in R&D processes entail extended control over later use and marketing of the result in order to obtain a reasonable rate of return. IPRs provide the means of protecting informational results of the production process. The increasing importance of the private sector, combined with new R&D technologies, thus calls for an expansion of IPRs. According to Swanson and Göschl, this expansion of IPRs leads in turn to growth in private sector R&D.

It becomes apparent from the above that, in discussing the interface of traditional knowledge and intellectual property rights within the context of modern biotechnology's impact on biodiversity, we find ourselves in the midst of a complex factual background characterized by a great variety of stakeholders and interests involved (Biber-Klemm, 2000a).

The question is whether legal instruments can be created which help balance inequities between formal and informal innovation processes, thus leading to a win-win situation for all stakeholders. The thesis is that incentives could be established to maintain traditional crops and knowledge by creating legal means to allocate their value.

[1]Traditional knowledge frequently lacks the criteria of novelty, the necessary inventive step, or the level of creativity seen as prerequisites for protection by industrial IPRs. Landraces as developed by informal breeding systems of farming communities fail to fulfil the criteria of stability and uniformity and thus fail to meet protective criteria of plant breeders' rights.

Aspects of the legal background

From the point of view of international public law, we find ourselves at the interface of the Convention on Biological Diversity (CBD), the International Undertaking on Plant Genetic Resources (IU) of the Food and Agriculture Organization (FAO)[2], the International Convention for the Protection of new Varieties of Plants (UPOV-Convention), and the Agreement on Trade-Related Aspects of Intellectual Property Rights (TRIPS).

Within the framework of these instruments and the communities involved, the legal means to regulate access to and use of traditional knowledge and genetic resources are intensively discussed. Emerging regional and national legislation seeks to resolve the antagonism between granting protection and facilitating access to information.

Intellectual property rights: TRIPS and UPOV

Property rights to protect informational values of human creativity and innovation are a common legal instrument in industrialised countries. IPRs were designed to balance the interests of the inventor in amortising his investments and the interest of the public in having access to the invention. The solution is a system of ongoing "inventive steps". IPRs confer to their owner the exclusive right to his creation for a limited time, but oblige him or her to disclose the invention. From an economic point of view, IPR systems are mechanisms that award extended rights of control over the marketing of certain goods in order to provide incentives for investing in generation of information (Swanson and Göschl, 2000).

Different forms of IPRs exist.[3] Patents and plant breeder rights are at the core of discussions on conflicting interests between IPRs and traditional knowledge.

Patents

A patent is a grant issued by a national government conferring the right to exclude others from making, using, or selling the invention within the national territory. Patents are generally defined to include scientific inventions concerning products and processes in all fields of technology, including machines, manufacturing products, chemical and electrical structures and compositions and processes. To be patentable, inventions have to be new, to involve an inventive step or, synonymous to be non-obvious and useful (Art. 27.1 TRIPS). Patent applicants must disclose their inventions "in a manner sufficiently clear and complete for the invention to be carried out by a person skilled in the art" (Art. 29.1 TRIPS).

Plant breeders' rights

Plant breeders' rights are intellectual property rights specifically designed to protect modern plant varieties as the result of formal breeding processes. They were established by the International Convention for the Protection of New Varieties of Plants (UPOV-Convention), which was concluded in 1961 and has since been revised four times. The protective criteria of plant breeders' rights require that a plant variety be new, distinct, homogenous (uniform), and stable. In particular, the last two criteria exclude landraces and traditional varieties from protection.[4]

---

[2] In the time between finalising the article and its publication, the IU has been replaced by the International Treaty on Plant Genetic Resources for Food and Agriculture, Rome, 3 Nov. 2001. See below footnote 8.

[3] The five basic types are patents, copyrights, trademarks, trade secrets, and industrial designs. Some of them are discussed in view of the potential to grant protection to informational values created by indigenous and local communities, e.g., Blakeney in WIPO 99/3.

[4] For details, see Girsberger, 1999, 156 ff.

Discussion
During recent decades the system of formal intellectual property rights has expanded continuously in different ways. New types of IPRs have been created (e.g., the described plant breeders' rights or the IPR to protect Integrated Circuits).[5] The system has been harmonised at the global level, the last step in this evolution being conclusion of the TRIPS Agreement. The scope of rights has been expanded, e.g., by acceptance of broad claims and narrow interpretation of restrictions on patentability (Drahos, 1999). This evolution causes specific problems in the context of patenting plant varieties. Broad claims can lead to monopolization of entire varieties and thus limit research and breeding activities (Correa, 1999; De Miranda Santos and Lewontin, 1997).

On the other hand, particularly in connection with the problem of patenting plant varieties, emerging discussions question the uniformity of the traditional concept of intellectual property rights, in particular for patents. According to Thurow (1997), "trying to squeeze today's developments into yesterday's systems of intellectual property rights simply won't work." He submits creation of a differentiated system corresponding to different industries, to the various types of knowledge and inventors. Reichmann (1998) pleads for development of a subpatentable right in order to regulate investment in "small grain-sized innovation that eludes the regulatory grasp of classical intellectual property laws." He argues that "unless a new market-driven liability regime... is developed to regulate investment in purely incremental, small grain-sized innovation, the prospects are for a dangerously overprotective proliferation of misbundled property rights that will spread like kudzu and strangle today's most vibrant creative forces, especially the small and medium-sized firms...."

Traditional knowledge and farmers' rights

Definition
Traditional knowledge – knowledge, innovation, and practices of indigenous and local communities embodying traditional lifestyles (Art. 8 (j) CBD) – can be characterized as a body of knowledge built by a group of people, indigenous peoples or farming communities, through generations of living in close contact with nature. It encompasses most fields of technology, ranging from traditional medicinal and agricultural practices to music, design, and the graphic and plastic arts. It is often integrated in a holistic view of the world and closely linked to the cultural, spiritual, and religious life of the community.

Traditional knowledge is in a continuous process of creation, development, and refinement and thus in permanent evolution. Mutual exchange plays a crucial role in this process. Traditional knowledge is often considered to be an achievement of the entire community. Its attribution, use, and tradition is frequently regulated by the customary law of communities[6].

Traditional knowledge can be in addition to a genetic resource, as is the case, e.g., with knowledge on a specific quality of a biologic resource, e.g., a pharmaceutical effect of a wild or domesticated medicinal plant or a crop variety's specific resistance against environmental impacts. However, traditional knowledge is also integrated into the genetic programme of cultivated varieties, which represent the experience and skill of generations of traditional breeders in adapting the variety to specific local circumstances.

As mentioned above, both types of traditional knowledge represent so-called informational goods. This fact is particularly obvious in the case of seeds of domesticated crops that are self-propagating and traded as foodstuff, the genetic information being enclosed in each grain.

[5] Washington Treaty on Intellectual Property in Respect of Integrated Circuits, 1989; Art. 35 to 38, TRIPS.
[6] For more details on the terminology, see Girsberger in this volume and the overview given by Blakeney in WIPO, 1999. Compare the summary of the results of the WIPO fact-finding missions in WIPO, 1999b, p. 3.

Existing intellectual property rights frequently cannot protect traditional knowledge. From the point of view of formal intellectual property rights, traditional knowledge is thus considered to be in the public domain if it is not kept secret, notwithstanding the system of customary law covering its use within the communities.

In the framework of international public law, regulations covering traditional knowledge follow different patterns from the institutional point of view: knowledge regarding plant genetic resources for food and agriculture (PGRFA) is contained in the concept of farmers' rights in the – non-binding – International Undertaking on Plant Genetic Resources (IU)[7]; traditional knowledge of indigenous and local communities in general, mainly encompassing additional knowledge, is taken up by the Convention on Biological Diversity (CBD).

The CBD

The CBD, which was concluded in 1992 at the occasion of the United Nations Conference on Environment and Development, addresses the issues described above: It reflects the scientific state of the art in integrating genetic diversity and genetic resources in its definition of biodiversity. It links conservation of biodiversity with economic issues and addresses both conservation and sustainable use of biological resources.

Of primordial importance in our context are regulations on access to genetic resources, on sharing benefits from their use, and on acknowledgement of traditional knowledge.

The CBD's regulation on access and benefit sharing establishes the basis for contractual relations between donor countries and prospecting firms. It prescribes that access to genetic resources be granted under the condition of prior informed consent, mutually agreed terms, and fair and equitable sharing of research and development results as well as benefits arising from use of genetic resources.

The existence, value, and importance of traditional knowledge is implicitly recognized by the CBD. The contracting parties are obliged to "respect, preserve and maintain" this knowledge and to promote its wider application with the approval and involvement of the holders of this knowledge. They are also to "encourage the equitable sharing of the benefits arising from the utilization of such knowledge, innovations and practices" (Art. 8 j).

However, these provisions are reserved to national legislation; therefore they confer no rights to individuals or groups but oblige the states to legislate in this matter "if possible and appropriate."

The International Undertaking and the farmers' rights

As in affects farming systems, the concept of traditional knowledge is closely linked to the notion of farmers' rights, crop varieties being the result of the breeding skill and knowledge of generations of farmers.

Farmers' rights are regulated in the IU, and there defined as "rights arising from the past, present and future contributions of farmers in conserving, improving, and making available plant genetic resources, particularly those in the centres of origin/diversity" (IU, Conference Resolution 5/89).

Creation of farmers' rights has been a reaction to balance expansion of plant breeders rights. However, their implementation still remains largely open. The IU is presently under revision. Two options to implement farmers' rights are discussed (Bragdon and Downes, 1998): 1) the "political" concept, which understands the notion of farmers' rights as a political slogan, as opposed to plant-breeders rights. In this concept, an international funding system to finance conservation and development of PGRFA is proposed (e.g., Girsberger, 1999; Brush, 1994); 2). The "legal" concept, which has been reinforced by conclusion of the CBD, encompasses

---

[7] See above footnote 2 and below, footnote 8.

proposals to define farmers' rights as an alternative form of IPR (e.g., Swaminathan, 1997; Shiva, 1999)[8].

Discussion
The input of traditional knowledge and landraces as conserved by local farmers into the industrial R&D process can be economically substantial (e.g., Ten Kate, 1999; Swanson and Göschl, 2000). According to the Swanson and Göschl analysis, investments effected at the basis of the respective production processes consist in maintaining a relatively greater diversity of genetic resources and dedicating greater amounts of land to retaining that diversity. The next investment step consists in observation and discriminatory use of genetic resources by local farmers and indigenous peoples. Time, knowledge, and resources are invested in producing information on effects of a plant or in selecting those genetic resources revealed by nature to be of greater fitness.

Whereas IPRs create incentives to invest in R&D in the following steps of formal R&D and industrial production, no comparable incentives exist at the level of earlier production stages, i.e., where basic information is created by informal R&D processes as effectuated by traditional breeders and indigenous peoples.

Swanson and Göschl (2000) conclude that, in order to maintain basic investments and innovation processes, incentives should also be created at the level where the information is created, i.e., at the level of the indigenous and farming communities.

Case studies

The following two case studies concerning 1) "piracy" of a South American landrace, and 2) patenting of the chemical compound extracted from a traditionally used tree root, illustrate problems described in the area of traditional knowledge related to seeds and pharmaceutical compounds. They help to clarify the issues that ought to be answered by legal means.

The Mexican yellow beans

Case
According to a report by the Rural Advancement Foundation International (RAFI) (2000a), in 1994 the owner of a small seed company and president of POD-NERIS L.L.C. bought a bag of commercial bean seeds in Sonora, Mexico, and took them back to the USA. He picked out the yellow-coloured beans, planted them, and allowed them to self-pollinate. He selected yellow seeds for several generations until he got what he describes as a "uniform and stable population" of yellow bean seeds.

He applied for a US patent on 15 November 1996, barely two years after he purchased the yellow beans. On 13 April 1999 he won an US patent for the "Enola" bean variety. The patent claims exclusive monopoly on any *Phaseolus vulgaris* (dry bean) having a seed colour of a particular shade of yellow. He claims that it is illegal for anyone to buy, sell, offer for sale, make use for any purpose including drying, eating, propagating, or importing yellow *Phaseolus vulgaris* of that description. On 28 May 1999 he won a US plant variety protection certificate

[8] The Treaty on Plant Genetic Resources for Food and Agriculture (above footnote 2) now regulates Farmers' Rights in its Article 9. There, the responsibility for realizing Farmers' Rights is delegated to the national governments. The following elements are in particular proposed to be included: protection of traditional knowledge, the right to equitably sharing the benefits arising from the utilization of PGRFA and the right to participate in the decision-making processes in related matters.

for the Enola bean variety. In late 1999, he brought legal suit against two companies that sell Mexican beans in the USA, charging that they infringe his patent monopoly. He demands royalties of six cents per pound on the yellow beans entering the USA from Mexico.

Yellow beans have been widely grown for centuries in Mexico.

Analysis and questions

The information patented in this case was obviously not novel but contained in an ancient Mexican or even South American landrace. It was situated in the public domain and freely accessible. In the context with patenting, the existence of the landrace ought to be considered as "prior art".

The following problems arise in this context: first the problem of proving the existence of the prior art. Patent examiners in the USA are not required to accept evidence of traditional knowledge held outside the USA as prior art, unless it has been reported (and thereby validated) by scientists and published in learned journals or otherwise made available to the public (Dutfield, 2000). In the case of the yellow bean, the proof was relatively easy, as the (genetically identical) variety of local beans was held by the "International Centre for Tropical Agriculture" as "in-trust" material. Under a 1994 agreement between the Consultative Group on International Agricultural Research (CGIAR) and the FAO, "in-trust" germ plasm is maintained in the public domain and cannot be included in any intellectual property claim.[9] However, proof of prior art would not be possible in case of knowledge handed down orally – as is the rule with traditional knowledge – or of plant varieties that are not documented in a public seeds bank.

Another question is how the original owner or author of the information can control and challenge abuse of the information in foreign patenting procedures, a prerequisite usually necessary to abolish a wrongly issued patent. This not only creates problems regarding access to the pertinent information but also know-how and financial means necessary to challenge a patent.

African tree root against fungal infections

Case

According to an article by Connor (2000) published in "The Independent" in February of this year, a research team from an European university has discovered a substance which has strong antimicrobial, specifically antifungal effects in the roots of an African tree (*Swartzia madagascariensis*). Laboratory tests have demonstrated that the compound is much more active than all existing compounds on the market. It is presently tested for treating fungal infections of the eyes, as there is almost no drug for treating eye infections from which many AIDS patients suffer. Traditional medicine describes use of preparations including the root of the tree to treat leprosy and syphilis as well as to kill termites.

The research team has just gained a patent on the substance in the United States.[10] An American company is developing it. It is hoped that it can be put on the market in a few years. Also it should soon be possible to synthesise the active ingredients in the laboratory. Connor mentions that "under the terms of an international treaty on biodiversity, a small percentage of the profits from new drugs extracted from the Third World has to be ploughed back into research within the country of origin." In this context, he cites the head of the research team stating that they "... want to give back to the traditional healers the information we get from studying the plant."

[9]This example was not a unique case of patenting in-trust material. See RAFI, 1998 and 2000a.

[10]US 5929124: Antimicrobial diterpenes.

Analysis and questions

This example demonstrates the process of developing a new product on the basis of traditional knowledge. On the side of the formal R&D process, additional (inventive) steps are to be made and funds must be invested until the product can be brought on the market. However, it is evident that the active substance would not have been discovered and the product would not have been developed without basic input by the knowledge of traditional healers. This leads to the following questions:

1. Is it – under the formula of fair and equitable sharing of benefits – sufficient to return information on the results of scientific analysis to traditional healers?
2. Who benefits from the traditional knowledge; i.e. who is the party obligated to share the benefits? The researchers in licensing the patent to a company? The company planning to produce and market the product? Or the company that will finally sell the synthesized product?
3. How can the amount of benefits be established and verified? How is the "fair share" of benefits to be determined? And in what form is this benefit to be shared? And with whom?[11]

Problems

The two examples, which could be supplemented by many others, amply demonstrate the complexity of questions emerging at the interface between traditional knowledge and intellectual property (rights). Currently, the unclear and highly politicised situation on the side of the states and peoples providing genetic resources and traditional knowledge as well as of intellectual property rights, fosters confusion and mistrust, leading to reluctance of countries and individuals or communities to grant access.

Thus in a growing number of cases indigenous populations deny research teams access, as was the case, e.g., in Chiapas, where 11 indigenous peoples' organizations demanded that a bioprospecting team funded by the US government suspend its activities (RAFI, 1999), and with the Wapishana Indians from Guyana, who intend to challenge two patents on compounds originating from a tree and bush grown locally that is used in traditional medicine and fishing. The challengers "will meet to discuss strategies for action and alert the other peoples about the need to preserve their knowledge." Meanwhile, some ancient rules of hospitality are revoked, and the entry of researchers is forbidden (Villamea and Pinto, 2000).

Grajal (1999), evaluating the situation from the angle of a researcher in natural science and in biodiversity, states that "under the magnifying glass of the new legislative frameworks, government agencies that traditionally process biodiversity research permits have adopted a "preventive mode", resulting in arbitrary moratoriums on export or collections permits for national or international museums or herbaria and in a general slowdown in biodiversity research permits." He states that "throughout the Andean region, publication of ethnobotany studies has stalled, pending clear guidelines on indigenous intellectual property rights on "intangible components" of access to genetic resources."

In the area of seeds, where cooperation and exchange have always been a crucial factor in the process of plant breeding, according to an analysis of the Swiss agency for agriculture (1999), this cooperation is threatened, due to conflicting interests between the industrialized and developing countries. On the side of the formal research and breeding, expansion of intellectual property rights of formal breeders lead to limit access to genetic information (Correa, 1999). According to an informal survey on cooperation between public and private

[11] This question is of course closely linked to the question as to who is to give the prior informed consent to the industrial development of the information in the first place.

plant breeding (Price, 2000), 48% of public breeders had experienced difficulty in obtaining genetic stocks from private companies; 45% indicated that this had interfered with their research; 28% felt that it had interfered with their ability to release new varieties.

The situation described clearly indicates the necessity of finding legal solutions in order to create a clear basis, transparency, and mutual trust for trade in informational values between the two worlds of the informal and formal innovation systems.

Elements for a legal solution

Analysis of case studies and factual background leads to the elements a legal solution ought to contain, namely: to spell out conditions and procedures for access to genetic resources including prior informed consent. This element is closely linked to the need to designate the parties to access- and benefit-sharing negotiations, in particular in view of prior informed consent and mutually agreed terms. Another element is the definition of benefits. Finally a solution must be found to control use of genetic resources and related traditional knowledge.[12]

With view to access and prior informed consent, competent authorities and entities are to be set up to negotiate mutually agreed terms. A balance must be found between the interests of:

1. national and international research, trade, and industry on easy access to information, a clear and concise approach to access negotiations, and timely issuance of research permits, and
2. the indigenous and farming communities to decide in an informed way on access to and use of their knowledge and innovation.

With regard to benefit sharing, the parties competent to negotiate must be clearly defined, as must the obligated and the beneficiary party. Benefit-sharing conditions must be spelled out for use of traditional knowledge and the material time of the discharge of the obligation. With view to controlling use of genetic resources and traditional knowledge, the question is how illicit use of informally generated information can be controlled and sanctioned an important aspect of this being the proof of prior art.

The analysis thus indicates the following elements for a legal solution:

1. a sound legal basis for access to genetic resources ought to be created
2. responsible agencies named
3. procedures outlined;
4. the parties competent to give prior informed consent and to negotiate sharing of benefits are to be clearly defined; and
5. mechanisms to control implementation ought to be created.

Legislation of the Cartagena Agreement and of Peru as an example for a regional and national solution

At the national level, in many countries, legislation to implement the access- and benefit-sharing obligation of the CBD are presently being created. As one example of this legislation, this study examines Decision 391 of the Cartagena Agreement on a common system on access to genetic resources and Peru's draft legislation on traditional knowledge and access- and benefit-sharing. The analysis concentrates on the interface of traditional knowledge and intellectual property rights, taking into consideration the elements elaborated above.

---

[12]For details regarding the designing legal frameworks for the access to genetic resources, see Glowka, 1998.

Decision 391 of the Cartagena Agreement

In general

The Cartagena Agreement of 1969 (the Andean Subregional Integration Agreement) as revised by the Protocol of Trujillo of 1996, creates the Andean Community which aims at economic and social integration of its members in view of the creation of a Latin American Common Market. The members of the Andean Community are Venezuela, Colombia, Ecuador, Peru, and Bolivia.

Decision 391 of the Cartagena Agreement, as signed by the Commission in July 1996, establishes a common regime on access to genetic resources. The decision encompasses the genetic resources, their by-products[13], and the so-called "intangible components" (i.e., the traditional knowledge). Its purpose is to regulate access to genetic resources of the Member Countries in order to 1) establish conditions for just and equitable participation in the benefits of access; 2) lay the foundations to recognize and evaluate genetic resources, their by-products and associated intangible components, 3) to promote conservation and sustainable use of biological diversity; 4) to promote development of scientific, technological and technical capacities at all levels, and 5) to strengthen the negotiating capacity of member countries (Art. 2). The decision submits genetic resources of every kind, i.e., wild and domesticated resources and their by-products, as parts of the heritage of each state to its sovereignty.

Traditional knowledge

The Decision defines the "intangible components associated to genetic resources" as "all know-how, innovation or individual or collective practice, with a real or potential value, that is associated with the genetic resource, its by-products or the biological resource that contains them, whether or not protected by intellectual property regimes" (Art. 1, para. 18).

In its preamble the decision explicitly recognizes the "strategic value" of the know-how, innovations, and practices of native Afro-American and local communities", the "historic contribution made by the native Afro-American, and local communities to the biological diversity, its conservation and development and the sustained use of its components" (para. 5 and 6) and acknowledges the close interdependence between native populations and communities and biological resources and the necessity of its reinforcement (para. 7).

According to the decision, the traditional knowledge remains under the authority of its authors and member states are obliged to recognize and value the rights and the authority of the communities to decide about its utilisation (Art. 7). Thus, the access contracts are to include conditions which aim at "strengthening and development of the capacities of the native, Afro-American and local communities with relation to the associated intangible components, the genetic resources and their by-products" (Art. 17f; general aspects of the access procedure): the supplier of traditional knowledge has to be identified in the application for access (Art. 26 b). In an appendix integrated into the access contract, fair and equitable distribution of profits from use of knowledge is to be stipulated and signed by the supplier of knowledge and the applicant for access.

In case of failure to comply with appendix stipulations, the entire access contract can be nullified (Art. 33). National IPR offices have to check compliance with access regulations. Access without authorization and transactions unprotected by corresponding contracts can be sanctioned by fines, confiscation, closing establishments, and disqualification from reapplying for access (Art. 46, 47). Intellectual property rights obtained or developed on the basis of illicit access are not to be acknowledged by member countries.

[13]By-products (Producto derivado) are defined as "a molecule, a combination or mixture of natural molecules, including crude extracts of live or dead organisms of biological origin that come from the metabolism of living beings" (Art. 1, para. 7).

The draft Peruvian system

Based on the Andean framework, Peru has drafted a regulatory instrument on access to genetic resources and on access to traditional knowledge: the "Proposal of a Regulation on Access to the Genetic Resources" (access) and the "Proposal of a Regime of the Collective Knowledge of the Indigenous Peoples" (TK).[14] [15]

The proposed regulation on access to genetic resources fixes the content of the request for access, determines the proper authorities, and fixes the procedure and period of time granted to the authorities for the procedure's individual steps. It spells out the minimum conditions to be contained in agreements. As parties to the negotiations with the purchaser/applicant, it determines a state agency and the provider of the genetic resource. The provider can be the state, but also, e.g., an ex-situ gene-bank or an indigenous people. The conditions of compensation are to be negotiated with the resource provider (Art. 20 k) – the state receiving a certain percentage of the transaction value agreed between the provider and applicant (Art. 22). Access to genetic resources located in lands of indigenous peoples must have their express consent (Art. 6).

The draft regime to protect knowledge of indigenous peoples creates a collective knowledge that belongs to the communitie*s* of indigenous peoples and – explicitly – not to its individual members (Art. 9). This collective knowledge is considered part of the indigenous peoples' cultural heritage. The present generations preserve, develop, and administer their collective knowledge in their own interest and in that of future generations. Therefore, the rights of the indigenous peoples to their collective knowledge are inalienable and not extinguishable (Art. 8). As a consequence, knowledge can only be licensed. The license does not prevent others from using the knowledge nor from other licenses being awarded on the same knowledge. Nor do licenses affect the right of peoples to use and further develop collective knowledge (Art. 25).

The draft regime to protect knowledge of indigenous peoples creates three instruments: a system to control access to the knowledge, a register of collective knowledge, and an additional funding system.

To gain access to traditional knowledge, the consent of the people *concerned* is needed; *i.e.*, not all people having the same knowledge must grant approval for access. Regarding the access procedure, a difference is made between access for scientific and commercial use. Only the prior informed consent is needed for access involving scientific use. In case of access for commercial or industrial application, an additional licence contract must be concluded. This contract must fix conditions on payment for access and equitable distribution of benefits (Art. 7). Furthermore, the licensing contract must contain the obligation to inform the licenser periodically and in general terms about advances in investigation, industrialisation, and marketing of products developed from collective knowledge (Art. 21 e).

The funding system is meant to support development of indigenous peoples. The fund is endowed in part according to percentages of economic benefits resulting from marketing goods developed on the basis of traditional knowledge.

The Register of Collective Knowledge seeks to

1. preserve collective knowledge,
2. identify collective knowledge, and
3. supply information on uses that indigenous peoples give to biological resources.

[14]The Proposal on Traditional Knowledge is based on a provision in the Peruvian law on intellectual property that gives the Ministry of Industry, Tourism, Integration, and International Trade the authority to establish a system to protect and if possible register the knowledge of native and farming communities.

[15]In the time between finalising the article and its publication the proposal has been adopted and promulgated as "Ley 27811: Régimen de protección de los conocimientos colectivos de los pueblos indígenas vinculados a los recursos biológicos" (10 august 2002).

The register has an optional and not a constitutive character. This means that all collective knowledge as defined by the system is protected, notwithstanding its registration.

Presentation of the licensing contract is a prerequisite to obtain a license of invention or a certificate of holder of vegetal variety (additional provisions).

Comments

Analysis of the Andean regional and Peruvian draft law regulations leads to the following conclusions: The draft law spells out conditions for *access* to traditional knowledge, parties involved in negotiations, and requirements for prior informed consent and mutually agreed terms. Indigenous peoples are integrated as providers of genetic resources situated in their lands and as holders of their collective knowledge. The obligatory involvement of the authorized state agency in the negotiations may alleviate inequalities in negotiating skills between the applicant and the indigenous peoples.

Integration of indigenous peoples in access negotiations complicates procedures from the point of view of formal R&D. Yet this effect is mitigated by the requirement that indigenous peoples must decide on a system of representation for the negotiations (Art. 13, TK). Deadlines fixed for response to the request may prevent long and delayed procedures. The differing procedures for access to traditional knowledge for scientific and commercial purpose ought to facilitate scientific research. Contracts on global access may serve the same end by granting the privilege of access to universities and centres of investigations with residence in the country (Art. 4, Access). But a difficulty resides in the difference between scientific research and research for industrial use with regard to traditional knowledge. What happens, e.g., if knowledge published in ethnopharmacological or ethnobotanical journals is used later in industrial research?

The registration mechanism might provide proof of prior art and authorship of the information. However, the criteria for registering traditional knowledge to fulfil this end ought to be researched and clearly defined.

To some extent, implementation of the provisions can be controlled at the national and regional level. Infractions of the provisions can be sanctioned under the system of decision 391 within states that are members of the Cartagena Agreement.

However, some important problems remain. Sanctioning of infractions outside the Andean Community is difficult. Only indirect measures are possible, since illicitly attributed property rights are not recognized within the Andean Community and further access can be denied. Thus abuse of information in the public domain (as was the case of the Mexican bean) remains particularly difficult to control. Furthermore benefit-sharing mechanisms, apart from up-front payments, are funded on mutual trust, as they require disclosure of the benefits and transparency with view to net gains.

It can be concluded from this that the scope to resolve problems at a national or regional level remains limited, given the global dimension of the interests involved and the international character of the markets. Viewed from this angle, problems become apparent in bringing about control at the international level and assuring fairness in sharing benefits. Further issues ask how shared benefits are obtained by those meriting them, and how mutual trust and confidence can be secured to assure further exchange of knowledge. Clearly these issues cannot be regulated at the national or regional level but call for international measures.

Proposed (legal) means on the international level

According to the criteria developed, the aims of instruments on the international level would be:
1. to control the legality of acquiring information,
2. to create positive incentives to conserve and maintain biodiversity and traditional knowledge;
3. to secure and maintain access for scientific research,
4. to support fair and equitable sharing of benefits, and
5. to favour research and industrialisation within the country of origin.
To this end, the following measures are discussed.

Code of conduct

In the framework of the follow-up process of the CBD, Switzerland has elaborated "Draft Guidelines on Access and Benefit Sharing Regarding the Utilisation of Genetic Resources" that were circulated at the last Conference of CBD parties in May 2000 (UNEP/CBD/COP/5/INF/21), but have not yet been officially discussed.[16]

Codes of conduct are voluntary instruments that spell out rules for business proceedings agreed upon by the stakeholders involved in transactions. Their object is to create a basis of transparency and mutual trust. The Draft Guidelines aim at providing a non-discriminatory framework for all stakeholders involved in access to genetic resources as well as fair and equitable sharing of benefits arising from their use (Art. 1). With view to traditional knowledge, the Guidelines stipulate an obligation of each user to respect customs, traditions, values, and customary laws of indigenous and local communities (Art. 7.2) and to assure that research and development activities in no way impede continuation of traditional use of genetic resources (Art. 7.5).

Creation of an international funding system

The aim of an international funding system would be to create financial incentives to maintain biological diversity and of traditional knowledge.

The option of creating an international funding system is discussed in the framework of the revised IU in the context of realising farmers' rights. The fund would permit compensating industrial use of traditional knowledge that cannot be allocated because determination of individual holders is difficult or impossible. This is frequently the case with information contained in PGRFA. The fund could be sponsored by industries using traditional PGRFA and by stakeholders having an interest in their conservation, i.e., by governments and formal plant breeders.

Adaptation of industrial intellectual property rights

On the side of the purchasing countries, measures are necessary to control lawful acquisition of plant genetic resources and traditional knowledge. Adaptation of industrial intellectual property rights is proposed to this end. The measures could consist, e.g., in:

[16]In the time between finalising the article and its publication, the proposed Code of Conduct has been revised and adopted by the Contracting Parties to the CBD as "Bonn Guidelines on Access to Genetic Resources and Fair and Equitable Sharing of the benefits Arising out of their Utilization" (Decision VI/24).

1. the obligation to disclose the origin of information and to certify its lawful acquisition in patent/plant breeders' rights application procedures,
2. agreement to withhold patents on subject matter available to the public by means of use or written description prior to the date of patent application filing, and
3. agreement that patents must not be granted on original material obtained from ex-situ collections or for naturally occurring biological material (discoveries).

Community traditional intellectual property rights

As has been shown above, the system of the industrial intellectual property rights has been adapted to new situations in recent decades, and discussions on differentiating within the IPR system in the industrial context are emerging. This evolution supports the claim to create a new type (or types) of property rights specially adapted to protect the collective traditional knowledge of indigenous peoples and farming communities at the national level but also internationally. The option of such a community "traditional intellectual property right" (TIP right) could, e.g., be integrated into the TRIPS agreement (e.g., Cottier, 1998).

Conclusion

It becomes clear from the above that no simple and uniform solution is apt to resolve the problems and questions at the interface of traditional knowledge and formally generated intellectual property. A variety of instruments will be needed, both at the national and international level.

In this context it must be stressed that valuation of traditional knowledge is not only an imperative with view to its maintenance and the maintenance of diversity. The claim is also to be seen as a way to balance equities between formal and informal innovation systems and to grant communities autonomy in deciding on the use made of their innovations.

References

Agreement on Trade-Related Aspects of Intellectual Property Rights, adopted Dec. 15, 1993 (TRIPS Agreement), reprinted in 33 I.L.M. 1197 (1994)

Andean Community, Decision 406 as approved by the Commission on 25 June 1997 on the Codification of the Andean Subregional Integration Agreement (Cartagena Agreement); http://www.comunidadandina.org/english/andean/ande_trie1.htm.)

Andean Community, Decision 391 as approved by the Commission on 2 July 1996 on the Common Regime on Access to Genetic Resources. (http://www.comunidadandina.org/english/Dec/d391e.htm)

Biber-Klemm S (2000a) Biotechnology and traditional knowledge – legal means to balance the equities. In: *International Journal of Biotechnology 2(1/2/3) 85-102*

Biber-Klemm S (2000b) Intellectual Property, Genetic Engineering, and Sustainable Development: Incentives to Bring about Conservation and Sustainable Use of Genetic Resources in the Framework of the World Trade Order. In: T Cottier, Mavroidis PC (eds) *Intellectual Property: Trade, Competition and Sustainable Development.* Proceedings of the World Trade Forum on Intellectual Property, Berne, August 17-28, 1999, Michigan University Press (in press)

Biber-Klemm S (2000c) The management of genetic diversity in agroecosystems: The role and function of law to conserve genetic diversity and to support its community based management. In: IMM Engels, Ramanatha Rao V, Brown AHD, Jackson MT (eds.): *Managing Plant Genetic Diversity. Oxon, New York, CABI Publishing*

Blakeney M (1999) What is traditional Knowledge? Why should it be protected? Who should protect it? For whom? Understanding the value chain. In: World Intellectual Property Organization (WIPO) Roundtable on Intellectual Property and Traditional Knowledge, WIPO/IPTK/RT/99/3

Bragdon SH, Downes DR (1998) Recent policy trends and developments related to the conservation, use and development of genetic resources. IPGRI: In: J Engels (ed.), *Issues in Genetic Resources* No. 7

Brush SB (1994) Providing Farmers' Rights through in-situ conservation of crop genetic resources. 44 P. Commission on Plant Genetic Resources, First Extraordinary Session, Rome 1994. *Background Study Paper* No. 3

Connor S. (2000) African Root could be Cure for Athlete's Foot *The Independent,* London, February 21

Convention on Biological Diversity, adopted June 5, 1992 (Biodiversity Convention, CBD), reprinted in 31 I.L.M. 818 (1992)

Convention on Biological Diversity (2000) Conference of the Parties. Access to Genetic Resources: Draft *Guidelines on Access and Benefit-Sharing regarding the Utilization of Genetic Resources* (UNEP/CBD/COP/5/INF/21, 3 April 2000)

Cottier T (1998) The Protection of Genetic resources and Traditional Knowledge: Towards more Specific Rights and Obligations. *Journal of International Economic Law* 555-584

Correa CM (1999) Access to plant genetic resources and intellectual property rights. Commission on Genetic Resources for Food and Agriculture, *Background Paper* No. 8

De Miranda SM, Lewontin RC (1997) Genetics, plant breeding and patents: conceptual contradictions and practical problems in protecting biological innovations. *Plant Genetic Resources Newsletter* 112:1-8

Drahos P (1999) Biotechnology, Patents, Markets and Morality. *European Intellectual Property Review*. 21 (9): 441-449

Dutfield G (2000) Intellectual Property Rights, *Trade and Biodiversity*, IUCN, Earthscan, London

Girsberger MA (1999) Biodiversity and the Concept of Farmers' Rights in International Law: Factual Background and Legal Analysis. Bern, Berlin etc., Peter Lang

Glowka L (1998) A Guide to Designing Legal Frameworks to Determine Access to Genetic Resources. In: IUCN (Ed): *Environmental Policy and Law Paper* No. 34

Grajal A (1999) Biodiversity and the Nation State: Regulating Access to Genetic Resources Limits Biodiversity research in Developing Countries. *Conservation Biology* 13 (1): 6-10

International Convention for the Protection of new Varieties of Plants of Dec. 2, 1961 , as revised at Geneva on 10 Nov. 1972 (UPOV Convention 1961), reprinted in UPOV Publication No. 295 and in The International Union for the Protection of New Varieties of Plants <www.upov.org/eng/convntns/1961/content.htm> (15 May 2000)

International Convention for the Protection of New Varieties of Plants of Dec. 2, 1961, as revised at Geneva on 10 Nov. 1972, on 23 Oct. 1978, and on 19 Mar. 1991 (UPOV Convention 1991), reprinted in UPOV Publication No 221 (E) and in The International Union for the Protection of new Varieties of Plants <www.upob.org/eng/convntns/1991/content.htm> (15 May 2000)

International Undertaking on Plant Genetic Resources, adopted at the 22nd Session of the FAO Conference as FAO Conference Resolution 8/83 on 23 Nov. 1983 (International Undertaking, IU), reprinted in International Undertaking on Plant Genetic Resources 1, FAO Doc. CPGR-Ex1/94/Inf.1 (Sept 1994)

International Undertaking on Plant Genetic Resources, Annex I, entitled "*Agreed Interpretation of the International Undertaking*" adopted as FAO Conference Resolution 4/89 at the 25th session of the FAO Conference on 29 Nov. 1989 (Annex I), reprinted in International Undertaking on Plant Genetic Resources 7, FAO Doc. CPGR-Ex1/94/Inf.1 (Sept 1994)

International Undertaking on Plant Genetic Resources, Annex II, entitled *"Farmers' Rights"*, adopted as FAO Conference Resolution 5/89 at the 25th Session of the FAO Conference on 29 Nov. 1989 (Annex II), reprinted in International Undertaking on Plant Genetic resources 9, FAO Doc. CPGR-Ex1/94/Inf. 1 (Sept 1994)

Peru: *Propuesta de Regimen de Proteccion de los Conocimientos Colectivos de los Pueblos Indigenas* (on file with author)

Peru: *Propuesta de Reglamento sobre acceso a los Recursos Geneticos* (on file with author).

Price S (1999) Public and private plant breeding. *Nature Biotechnology* 17: 938

Reichmann JH (1998) *Solving the Green Tulip Problem: Packaging Rights in Subpatentable Innovation*, Draft June 18 (on file with author)

Rural Advancement Foundation International (RAFI) (1998) *Plant Breeders Wrongs*. (www.rafi.org., 29 May 2000)

Rural Advancement Foundation International (RAFI) (1999) *Biopiracy in Chiapas, Mexico. Denounced by Mayan Indigenous Groups*. (www.rafi.org/news; News Release, 1 December 1999)

Rural Advancement Foundation International (RAFI) (2000a) *Biopiracy – RAFI's Sixth Annual Update.* RAFI Communiqué, Issue 65

Rural Advancement Foundation International (RAFI) (2000b). *Mexican Bean Biopiracy. US-Mexico Legal Battle Erupts over Patented "Enola" Bean.* http://www.rafi.org, 17 January 2000

Shiva V (1999) Neem. A case study of intellectual property rights and traditional knowledge. Council of Europe (ed.): *International Conference on Ethical Issues arising from the Application of Biotechnology*. Oviedo, Spain. Doc. SPK/07/PROV/SUM/E

Swaminathan MS (1997) Ethics and equity in the collection and use of plant genetic resources: Some issues and approaches. In: International Plant Genetic Resources Institute (ed): *Ethics and equity in conservation and use of genetic resources for sustainable food security*. Proceedings of a workshop to develop guidelines for the CGIAR, Foz do Iguaçu, Brazil 1997, 7-18

Swanson T, Göschl T (2000) Property rights issues involving plant genetic resources: implications of ownership for economic efficiency. *Ecological Economics* 32: 75-92

Swanson T (1995) The appropriation of evolution's values: an institutional analysis of intellectual property regimes and biodiversity conservation. Swanson T. (Ed.) *Intellectual Property Rights and Biodiversity Conservation: an Interdisciplinary Analysis of the Values of Medicinal Plants.* Cambridge University Press. 141-175

Swiss Agency for Agriculture (1999) Bericht zum Stand der Verhandlungen der FAO im Bereich pflanzengenetischer Ressourcen für Ernährung und Landwirtschaft *(on file with author)*

Ten Kate K, Laird S (1999) *The Commercial Use of Biodiversity: Access to Genetic Resources and Benefit Sharing*. London, Earthscan

Thurow LC (1997) Needed: A New System of Intellectual Property Rights. *Harvard Business Review*. 95-103

Villamea L, Pinto M (2000) Biopiracy: Indians want patent. Chiefs prepare international Law suit against scientist who registered indigenous knowledge. *Grain*, BIO-IPR docserver, 19 January 2000

World Intellectual Property Organization (WIPO) (1999) Roundtable on Intellectual Property and Traditional Knowledge. In: M. Blakeney (ed.) *What is traditional Knowledge? Why should it be protected? Who should protect it? For whom? Understanding the value chain.* WIPO/IPTK/RT/99/3

World Intellectual Property Organization (WIPO) (1999b) *Protection of Traditional Knowledge: A Global Intellectual Property Issue*. Document prepared by the International Bureau. WIPO/IPTK/RT/99/2

Methods for Risk Assessment of Transgenic Plants
IV. Biodiversity and Biotechnology
K. Ammann, Y. Jacot and R. Braun (eds)
© 2003 Birkhäuser Verlag Basel/Switzerland

Intellectual property rights and traditional knowledge: background, terminology and issues arising

Martin A. Girsberger

Swiss Federal Institute of Intellectual Property, Legal Services, Patent and Design Law, Division of Legal and International Affairs, Einsteinstr. 2, CH – 3003 Bern, Switzerland

Introduction

In the past years, intellectual property rights (IPRs) and traditional knowledge (TK) have been the subject of debate in several international bodies, including the Convention on Biological Diversity (Biodiversity Convention),[1] the World Intellectual Property Organization (WIPO) and the World Trade Organization (WTO). Generally, this debate has been rather controversial, clearly showing the divergent views that exist on the two issues among developed and developing countries, private industry, indigenous and local communities, and non-governmental organizations (NGOs). Key words that have been frequently used in this debate are "green gold", "biopiracy", "exploitation of the South", "erosion of TK", "fair and equitable benefit sharing", "self-determination of indigenous peoples" and "no patents for life". Some of the debate has been conducted on an emotional level, rendering the debate even more difficult and hardening unnecessarily the positions of those involved. IPRs and TK are thus political "hot potatoes" that bear many pitfalls and are reason for controversy on the international level. The debate has also shown that at present a great deal of information is lacking on the two issues. The very definition of the term TK, for example, remains questionable.

Why have IPRs and TK become the subject of international debate? Why has this debate been so controversial? What is the relationship between IPRs and TK? And finally, what is their mutual relevance? This paper addresses these questions in three steps: First, it provides the background information necessary to understand the problematic area of IPRs and TK. Second, because clear terminology is one important means to avoid misunderstandings and is thus crucial for a fruitful debate, it defines the terms IPRs and TK. And third, it analyzes the proposal to protect TK by IPRs. Issues dealt with are the aims of the protection of TK, the systems available to protect TK and the suitability of IPRs for this protection.

The international debate has so far focused on the TK of indigenous and local communities, especially those living in developing countries. Even though other parts of the population may also hold TK,[2] their TK has hardly been an issue on the international level. Parallel to the international debate, this paper focuses on the TK of indigenous and local communities. Nevertheless, a large part of the arguments advanced and of the principles developed can also be applied to the TK of the other parts of the population.

---

[1] Reprinted in 31 I.L.M. 818 (1992).

[2] Examples of such TK are cures for diseases ("grandmother's remedies"), agricultural practices and environmental knowledge. A substantial part of this TK, however, has been replaced by modern knowledge and has consequently been lost.

Background

For centuries, TK has been used almost exclusively by its creators, that is, indigenous and local communities. Access to, use of and handing down of TK has been regulated by local laws, customs and traditions. Some TK has been freely accessible by all members of an indigenous or local community and has been freely exchanged with other communities; other TK has only been known to particular individuals within these communities such as shamans, and has been handed down only to particular individuals of the next generation. Over many generations, indigenous and local communities have accumulated a great deal of TK which has generally been adapted, developed and improved by the generations that followed.

For a long time, Western anthropologists and other scientists have generally been able to freely access TK and have documented it in their works. Still, this TK was only seldom used outside the indigenous and local communities that created it. More recently, however, Western scientists have become aware that TK is neither outdated nor valueless knowledge, but, instead, can be useful to solve some of the problems facing today's world.[3] Modern science, for example, has shown an increased interest in some forms of TK as knowledge that can be used in research and development (R&D) activities and be integrated in modern innovations.[4] This holds especially true for TK regarding genetic resources, which has been integrated in modern pharmaceuticals,[5] agro-chemicals and seed.[6] With the improvements that can be expected to be achieved in modern biotechnology, the interest in TK regarding genetic resources is likely to further increase in the future. Other forms of TK that can be of interest to modern science include TK regarding the sustainable management of natural resources or the taxonomy of plants and animals.

---

[3]See, e.g., FARMER FIRST. FARMER INNOVATION AND AGRICULTURAL RESEARCH at xix (Robert Chambers et al. eds., 1989); Martha Johnson, *Dene Traditional Knowledge,* last visited Dec. 20, 1999, <http://www.carc.org/pubs/v20no1/dene.htm>. *See generally* Lee P. Breckenridge, *Protection of Biological and Cultural Diversity: Emerging Recognition of Local Community Rights in Ecosystems Under International Environmental Law,* 59 TENN. L. REV. 735, 744-50 (1992) (describing the emerging recognition of the value of the cultural diversity of local communities as a global resource related to the conservation of biological diversity).

[4]For examples of the use of indigenous knowledge in Western innovations see Naomi Roht-Arriaza, *Of Seeds and Shamans: The Appropriation of the Scientific and Technical Knowledge of Indigenous and Local Communities,* 17 MICH. J. INT'L L. 919, 921-25 (1996).

[5]For example, D-tubocurarine, a drug used as a skeletal muscle relaxant, is derived from curare, an Amazonian arrow poison. (Sarah L. Timpson & Luis Gomez-Echeverri, *Foreword* to RURAL ADVANCEMENT FOUND. INT'L [RAFI], CONSERVING INDIGENOUS KNOWLEDGE: INTEGRATING TWO SYSTEMS OF INNOVATION at i, i [1994]).

[6]According to Brush, "[i]ndustrial countries who use biological resources from less developed countries rely on indigenous knowledge in three ways. First, biological resources such as diverse crop populations have been screened, selected, and maintained by hundreds of generations of farmers and plant gatherers, and they reflect the distilled experience of thousands of individual selections. Second, local knowledge systems of indigenous farmers are used directly in the collection of biological resources, since plant collectors from industrial countries often rely on indigenous informants and guides in their search for useful plants. Third, industrial countries depend on indigenous knowledge for the conservation of biological resources." (Stephen B. Brush, *Indigenous Knowledge of Biological Resources and Intellectual Property Rights: The Role of Anthropology*, 95 AM. ANTHROPOLOGIST 653, 660 [1993]).

According to the Executive Secretary of the Biodiversity Convention, "[a]ccess to genetic resources and benefit sharing arising out of the use of genetic resources cannot be separated from the traditional knowledge of indigenous and local communities. Traditional knowledge can be highly valuable in identifying sources of new products derived from genetic resources, including pharmaceuticals and crop varieties." (Reprinted in *Traditional Knowledge and Biological Diversity* para. 29, UNEP Doc. UNEP/CBD/TKBD/1/2 [Oct. 18, 1997]). The Executive Secretary of the Biodiversity Convention furthermore states that "[i]nnovations and knowledge of local and indigenous communities... [are] often inherent in the genetic resource as intangible component or only the information given by the community makes the resource interesting to users outside that community." (*Id.* para. 31). See generally, e.g., John Woodliffe, *Biodiversity and Indigenous Peoples*, *in* INTERNATIONAL LAW AND THE CONSERVATION OF BIOLOGICAL DIVERSITY 255, 255-69 (Michael Bowman & Catherine Redgwell eds., 1996).

The use of TK outside its traditional area of use raises new legal problems, as the local laws, customs and traditions are no longer adequate to regulate access to and use of the TK in question. Outside the scope of these local rules, TK generally belongs to the public domain. With this legal status, TK can be freely used by all, that is, it can be used without authorization or compensation. Some of the modern innovations containing TK have been protected by IPRs.[7] The holders of the used TK generally neither participate in the granted IPRs nor do they receive any of the benefits that arise from the use of their TK. The situation has been additionally fueled by highly publicized cases where TK or the related genetic resources have been protected by IPRs, obviously without any additional creative efforts of the holders of the granted IPRs.

TK is not only of increased interest to modern science, but also of great value and importance for a large part of the world's population[8]. Still, it is lost at an accelerating rate.[9] This loss is often caused by factors that lay beyond the sphere of influence of the holders of TK, including the following: First, some governments have expressly suppressed or destroyed TK.[10] Second, the destruction of the traditional living space of indigenous and local communities[11] prevents the holders of TK from living as they previously have or causes them to vanish altogether.[12] Third, the imposition of the Western concept of civilization[13] on indigenous and local communities has in many cases caused that young members of these communities lose interest in the further use and development of TK. Fourth, free use of TK and the related genetic resources means that those responsible for their creation and conservation are unable to benefit from its increased economic value, causing its continued loss.

Several solutions have been proposed to change the described situation, including

1. the application of existing forms of IPRs to TK,[14]
2. the creation of *sui generis* systems of IPRs specifically designed to protect TK,
3. the registration and documentation of TK in data bases, and
4. the indication of the source of the used TK in patent applications.

---

[7] The company WR Grace, for example, was granted two patents concerning the neem tree, even though this tree is widely used by indigenous peoples in Asia and Africa for medicinal and pesticidal purposes. (RURAL ADVANCEMENT FOUND. INT'L [RAFI], CONSERVING INDIGENOUS KNOWLEDGE: INTEGRATING TWO SYSTEMS OF INNOVATION 8 [1994]).

[8] For example, eighty percent of all people rely on TK for their medical needs, and at least fifty percent of all people depend on TK regarding plants, animals and farming systems for their nutrition. (*Id.* at 1).

[9] According to the Executive Secretary of the Biodiversity Convention, "[t]he rate of erosion of the traditional knowledge of biodiversity held by indigenous and local communities has never been so high as it is in the current generation." (Reprinted in *Traditional Knowledge and Biological Diversity* para. 61, UNEP Doc. UNEP/CBD/TKBD/1/2 [Oct. 18, 1997]).

The Rural Advancement Foundation International (RAFI) states that "[m]any in the international community believe that the extinction of indigenous knowledge is both inevitable and even desirable. Indigenous knowledge is sometimes regarded as a barrier to the transmittal of new technological tools and information. If indigenous knowledge once had a role, it is said, that role has since been overtaken by others." (RURAL ADVANCEMENT FOUND. INT'L [RAFI], CONSERVING INDIGENOUS KNOWLEDGE: INTEGRATING TWO SYSTEMS OF INNOVATION 1 [1994]).

[10] See, e.g., Mongane Wally Serote, *Initiatives for Protection of Rights of Holders of Traditional Knowledge, Indigenous Peoples and Local Communities*, document prepared for the WIPO Roundtable on Intellectual Property and Indigenous Peoples, Geneva, July 23-24, 1998, last visited Dec. 20, 1999 <http://www.wipo.int/eng/meetings/1998/indip/rt98_4c.htm>.

[11] About 15,000 culturally distinct ethnic communities have been identified. (RURAL ADVANCEMENT FOUND. INT'L [RAFI], CONSERVING INDIGENOUS KNOWLEDGE: INTEGRATING TWO SYSTEMS OF INNOVATION at vi [1994]).

[12] In Brazil, for example, 90 Indian tribes have vanished since the beginning of this century. Of the 180 tribes remaining today, two thirds consist of less than one thousand people. (EDWARD O. WILSON, THE DIVERSITY OF LIFE 322 [W. W. NORTON & CO. 1993] [1992]).

[13] Wilson states that "much of the indigenous knowledge is being lost as European culture continues to intrude and the last preliterate native cultures in tropical countries weaken and disappear." (*Id.* at 43-44).

[14] See generally: Martin A. Girsberger, *The Protection of Traditional Plant Genetic Resources for Food and Agriculture and the Related Know-How by Intellectual Property Rights in International Law—The Current Legal Environment*, 1 J. WORLD INTELL. PROP. 1017 (1998).

IPRs, as a form of private property rights, would remove TK from the public domain, preventing the free use of TK in R&D activities and modern innovations. IPRs are probably the strongest legal means for the protection of TK and have so far dominated the international debate. This solution, however, is vested with many problems. The very nature of TK as a form of intellectual property, for example, is rather questionable. It is thus hardly surprising that at present no internationally accepted solution involving IPRs exists. The registration and documentation of TK in data bases is seen as one means to promote the identification and accessibility of existing prior art,[15] the criterion decisive when determining the novelty of inventions. Finally, the indication of the source of the used TK or the related genetic resources in patent applications is seen as one means to promote the fair and equitable sharing of the benefits arising from the use of TK, as it would allow the tracking and identification of the TK or the related genetic resources integrated in modern innovations.

The proposed solutions are not only of different nature and purpose, but also have varying legal effects. Furthermore, all of the proposed solutions raise questions that need to be answered before TK can adequately be protected. Because these questions not only concern legal, but also scientific, economic and political issues, viable answers have been difficult to find. Beside academia,[16] several international bodies have become active regarding IPRs and TK, including the following:[17] First, several conferences held under the auspices of the Biodiversity Convention[18] dealt with the two issues. In November 1997, the "Workshop on Traditional Knowledge and Biological Diversity" was held in Madrid, Spain;[19] in October 1999, the "Panel of Experts on Access and Benefit-Sharing" was held in San José, Costa Rica;[20] and in March 2000, the "Ad Hoc Open-Ended Inter-Sessional Working Group on Article 8(j) and Related Provisions of the Convention on Biological Diversity" will meet for the first time in Seville, Spain. Second, the World Intellectual Property Organization (WIPO) included TK in its work program for the years 1998-1999 and 2000-2001. In 1998, WIPO conducted fact finding missions in several countries. Furthermore, WIPO organized the "Roundtable on Intellectual Property and Indigenous Peoples" and the "Roundtable on Intellectual Property and Traditional Knowledge", which were held in Geneva in July 1998 and November 1999, respectively. And third, during the preparations for the 3rd Ministerial Conference of the World Trade

---

[15] See generally *Report of the Panel of Experts on Access and Benefit-Sharing* para. 138, UNEP Doc. UNEP/CBD/COP/5/8 (Nov. 2, 1999).

[16] See generally, e.g., Robert N. Clinton, *The Rights of Indigenous Peoples as Collective Group Rights*, 32 ARIZ. L. REV. 739 (1990); James O. Odek, *Bio-Piracy: Creating Proprietary Rights in Plant Genetic Resources*, 2 J. INTELL. PROP. L. 141 (1994); Mark Ritchie et al., *Intellectual Property Rights and Biodiversity: The Industrialization of Natural Resources and Traditional Knowledge*, 11 ST. JOHN'S J. LEGAL COMMENT. 431 (1996); Naomi Roht-Arriaza, *Of Seeds and Shamans: The Appropriation of the Scientific and Technical Knowledge of Indigenous and Local Communities,* 17 MICH. J. INT'L L. 919 (1996).

[17] For information on the activities of other international bodies see *Traditional Knowledge and Biological Diversity* paras. 114-54, UNEP Doc. UNEP/CBD/TKBD/1/2 [Oct. 18, 1997]).

[18] The Biodiversity Convention was adopted at the Conference for the Adoption of the Agreed Text of the Convention on Biological Diversity, held in Nairobi in May 1992, and entered into force on December 29, 1993. (LYLE GLOWKA ET AL., A GUIDE TO THE CONVENTION ON BIOLOGICAL DIVERSITY 1 [1994]). The Biodiversity Convention contains several provisions on TK. The main provision is Article 8(j), which, in the context of *in situ* conservation, deals with the TK of indigenous and local communities. Article 8(j) states that "[e]ach Contracting Party shall, as far as possible and as appropriate:...(j) [s]ubject to its national legislation, respect, preserve and maintain knowledge, innovations and practices of indigenous and local communities embodying traditional lifestyles relevant for the conservation and sustainable use of biological diversity and promote their wider application with the approval and involvement of the holders of such knowledge, innovations and practices and encourage the equitable sharing of the benefits arising from the utilization of such knowledge, innovations and practices." (Reprinted in 31 I.L.M. 818 [1992]). Other relevant provisions of the Biodiversity Convention are Articles 10(c), 17(2) and 18(4).

[19] See generally *Report of the Workshop on Traditional Knowledge and Biological Diversity*, UNEP Doc. UNEP/CBD/TKBD/1/3 (Dec. 15, 1997).

[20] See generally *Report of the Panel of Experts on Access and Benefit-Sharing*, UNEP Doc. UNEP/CBD/COP/5/8 (Nov. 2, 1999).

Organization (WTO), several developing countries proposed to include TK in the agenda of this conference. This proposal, however, was generally opposed by developed countries. Up to now, however, these international bodies have been unable to come up with the necessary answers. Considering the difficult nature of their task, this comes as no surprise.

Until recently, the international community has shown little or no interest in the protection of TK. Several reasons, including the following, can be advanced why the international community and especially developed countries should become more actively involved in the realization of this protection. First, the lack of protection can be used by the developing countries as a powerful argument against the improved protection of the existing forms of IPRs. Realizing the protection of TK could prove to be one powerful means to strengthen mutual trust between developed and developing countries, as it would show an interest of developed countries in an important concern of developing countries, NGOs and indigenous and local communities. Second, having rules regulating access to and use of TK will provide legal certainty for Western scientists and the private industry interested in accessing TK and using it in their R&D activities. And third, depending on the form of protection chosen, the conservation, further development and sustainable use of TK will be encouraged. This will not only serve the interests of the developing countries, but also of the developed countries.

Terminology

Intellectual property rights

IPRs[21] are "rights protecting innovation, creativity, or goodwill created in the marketplace".[22] As a form of private property rights, they grant their holders the exclusive right to use and commercially exploit the protected intellectual good. IPRs thus create a monopoly in the hands of their holders that allows them to prevent third parties from carrying out these activities.

Some documents and oral statements made at international conferences request the protection of the IPRs of indigenous and local communities. At first sight, the use of the term IPRs in this context may seem appropriate, especially if one considers that modern knowledge – the counterpart of TK - can generally be protected by IPRs.

Nevertheless, it should be kept in mind that the term IPRs has a specific legal meaning. The question thus arises whether these documents and oral statements use the term IPRs in this legal sense or whether they use the term more broadly, that is, to describe the intellectual

---

[21]See generally, e.g., Stephen B. Brush, *Indigenous Knowledge of Biological Resources and Intellectual Property Rights: The Role of Anthropology*, 95 AM. ANTHROPOLOGIST 653, 654-57 (1993) (providing a brief overview of the development of intellectual property rights and discussing possible justifications for intellectual property rights); Thomas Cottier, The Protection of Intellectual Property Rights: A Requirement for Technology Cooperation, Foreign Investment and Equitable Returns in Biotechnology Prospecting, *in* BIOTECHNOLOGIE FÜR ENTWICKLUNGSLÄNDER? CHANCEN UND RISIKEN DER BIOTECHNOLOGIE BEI LANDWIRTSCHAFTLICHEN NUTZPFLANZEN 65, 65-72 (Schweizerisches Zentrum für Internationale Landwirtschaft [ZIL] ed., 1995); David Hurlbut, *Fixing the Biodiversity Convention: Toward a Special Protocol for Related Intellectual Property*, 34 NAT. RESOURCES J. 379, 382-87 (1994) (discussing the viewpoints on property existing in different societies); THE CRUCIBLE GROUP, PEOPLE, PLANTS, AND PATENTS: THE IMPACT OF INTELLECTUAL PROPERTY ON TRADE, PLANT BIODIVERSITY, AND RURAL SOCIETY 19-21 (1994) (discussing the effects of biotechnology on present intellectual property rights systems); *id.* at 55-73 (discussing patents, plant breeders' rights and *sui generis* protection systems).

[22]Thomas Cottier, The Protection of Intellectual Property Rights: A Requirement for Technology Cooperation, Foreign Investment and Equitable Returns in Biotechnology Prospecting, *in* BIOTECHNOLOGIE FÜR ENTWICKLUNGSLÄNDER? CHANCEN UND RISIKEN DER BIOTECHNOLOGIE BEI LANDWIRTSCHAFTLICHEN NUTZPFLANZEN 65, 65 (Schweizerisches Zentrum für Internationale Landwirtschaft [ZIL] ed., 1995).

achievements and heritage of indigenous and local communities in general.[23] In case the term IPRs is used in this broader sense, it is imperative to use alternative terms in order to avoid unnecessary misunderstandings. Furthermore, the use of alternative terms will avoid that the discussion on the protection of TK focuses from the beginning on IPRs as the principal means to provide this protection. Viable alternatives to the term IPRs are "intellectual achievements", "intellectual contributions" and "intellectual heritage".

Traditional knowledge

The term knowledge[24] can be defined as "the fact or condition of knowing something with familiarity gained through experience or association";[25] "acquaintance with or understanding of a science, art, or technique";[26] "the sum of what is known: the body of truth, information, and principles acquired by mankind";[27] or "the information held in human memories that is accessible, by recall and the practice of learned skills, in a useful way in day-to-day life. In the

[23]For example, this seems questionable for the following statements. According to Baer "[t]he present legal situation is the result of a grim unlawful past. It is hard to see how to find a lasting settlement without resolving core problems, one of which is the lack of protection for fundamental indigenous rights, such as intellectual property rights. Commercial interests very often violate indigenous intellectual property rights. Although, such violations often do not formally constitute a breach of written legal standards, as neither national legislations nor international standards acknowledge the rights of indigenous peoples, these enterprises are still accountable to indigenous customary law." (Lars A. Baer, Initiatives for the Protection of Rights of Holders of Traditional Knowledge, Indigenous Peoples and Local Communities, document prepared for the WIPO Roundtable on Intellectual Property and Indigenous Peoples, Geneva, July 23-24, 1998, last visited Dec. 20, 1999 <http://www.wipo.int/eng/ meetings/1998/indip/rt98_4a.htm>).
Brush states that "[t]he assertion that indigenous knowledge should merit protection as intellectual property can thus be read in different ways. One reading, stressing the rights of tribal people unwillingly enmeshed in a nation-state, would emphasize the use of intellectual property as a means to express ethnic autonomy and to redress exploitation. Another reading, stressing the merits of local knowledge, would emphasize the use of intellectual property to provide incentives for all people to experiment with and protect biological resources. The ambiguity in the meaning of *indigenous* raises the question of whether intellectual property is to be extended to only tribal people and ethnic minorities within nation-states or to a general type of knowledge, regardless of the political and cultural status of people who control it.... [T]his confusion poses a major obstacle to implementing the new form of intellectual property rights." (Stephen B. Brush, *Indigenous Knowledge of Biological Resources and Intellectual Property Rights: The Role of Anthropology*, 95 AM. ANTHROPOLOGIST 653, 660 [1993]).
The Executive Secretary of the Biodiversity Convention states that "[w]ith regard to Article 16 [of the Biodiversity Convention], while indigenous and local communities can contribute many useful technologies for the conservation and sustainable use of biodiversity, the field is rich for scientific and technological collaboration for the management of biodiversity. However, where indigenous and local community knowledge or technologies are involved, their intellectual property rights must be protected. Because current regimes require further strengthening, consideration must be given to a range of additional options, such as alternative legal means, the adaptation of existing systems and the development of *sui generis* systems...." (Reprinted in *Traditional Knowledge and Biological Diversity* para. 9, UNEP Doc. UNEP/CBD/TKBD/1/2 [Oct. 18, 1997]).
The Executive Secretary of the Biodiversity Convention furthermore states that "[i]n order to protect and encourage, the necessary conditions may be in place, namely, security of tenure over traditional terrestrial and marine estates; control over and use of traditional natural resources; and respect for the heritage, languages and cultures of indigenous and local communities, best evidenced by appropriate legislative protection (which includes protection of intellectual property, sacred places, and so on)." (*Id.* para. 99).

[24]The term knowledge has to be distinguished from the term know-how, which is the "knowledge of how to do something smoothly and efficiently." (*Know-how* [visited Jan. 5, 2000] <http://www.eb.co.uk:180/cgi-bin/dictionary?va=know-how>).

[25]*Knowledge* (visited Jan. 5, 2000) <http://www.eb.co.uk:180/cgi-bin/dictionary?va=knowledge>.

[26]*Id.*

[27]*Id.*

context of TK it is often used to mean wisdom, which implies a blend of knowledge and experience integrated with a coherent world view and value system."[28]

The qualification of knowledge as "traditional" refers to "knowledge that has been accumulated by societies in the course of long experience in a particular place, landscape or ecosystem",[29] and that is handed down "by word of mouth or by example from one generation to another without written instruction."[30] The qualification of knowledge as "traditional" thus refers to the nature and purpose of this knowledge and the manner of its creation, use and handing down. This qualification, however, neither implies that this knowledge is backward or primitive, nor does it intend to judge the economic, social, environmental or other value of this knowledge.

In literature, several forms of TK are distinguished. A more detailed analysis of the descriptions and definitions put forward, however, reveals a problem of terminology. On one hand, some forms of TK distinguished in different documents are very similar or at least closely related. Often, it is thus rather a difference in terminology than a difference in substance, that is, different terms are used to designate the same or at least similar forms of TK. On the other hand, terms such as "indigenous knowledge", "traditional indigenous knowledge", "elders' knowledge", "local knowledge" and "traditional ecological knowledge" are used interchangeably in the same document to designate one form of TK,[31] whereas other documents use these terms to designate differing forms of TK. Unclear terminology presents a considerable obstacle to the categorization of TK and complicates the comparison between the different forms distinguished in literature. Considering the problem of terminology and the great diversity of TK existing, any attempt to categorize TK is thus likely to remain artificial and vague: "There are many terms in use to describe the body of expertise and knowledge held in indigenous communities.... None is wholly adequate or satisfactory."[32]

The following overview contains some of the forms of TK distinguished in literature. Described in more detail are traditional ecological knowledge,[33] traditional indigenous knowledge, traditional pharmaceutical knowledge and traditional agricultural knowledge.

Traditional ecological knowledge[34]

This form of traditional knowledge can be subdivided into several sub-categories, including the following:

1. First, traditional forest-related knowledge, which consists of five closely linked features; these are:
   - information on the soils, trees and other plants, animals, water streams, hunting grounds, and sacred sites;
   - rules for using these components;

[28]Reprinted in *Traditional Related Knowledge and the Convention on Biological Diversity* para. 22, UNEP Doc. UNEP/CBD/SBSTTA/2/Inf.3 (July 19, 1996).

[29]*Id.*

[30]*Tradition* (visited Jan. 5, 2000) <http://www.eb.co.uk:180/cgi-bin/dictionary?va=traditional>.

[31]See, e.g., *Recommendations on the Integration of Two Ways of Knowing: Traditional Indigenous Knowledge and Scientific Knowledge* of the Seminar on the Documentation and Application of Indigenous Knowledge held in Inuvik, Northwest Territories, Canada, on Nov. 15-17, 1996, <http://www.inusiaat.com/tek.htm>, last visited Dec. 20, 1999.

[32]*Id.*

[33]The acronym TEK is not only used for the term traditional ecological knowledge, but also for the term traditional environmental knowledge. To avoid confusion, the acronym is not used in this paper.

[34]This term is used in *Traditional Knowledge and Biological Diversity* para. 165, UNEP Doc. UNEP/CBD/TKBD/1/2 (Oct. 18, 1997). The term traditional ecological knowledge is sometimes used as being synonymous with the term traditional indigenous knowledge. (*See, e.g., Recommendations on the Integration of Two Ways of Knowing: Traditional Indigenous Knowledge and Scientific Knowledge* of the Seminar on the Documentation and Application of Indigenous Knowledge held in Inuvik, Northwest Territories, Canada, on Nov. 15-17, 1996, <http://www.inusiaat.com/tek.htm>, last visited Dec. 20, 1999).

- relationships among the different various components of a particular forest ecosystem, including information on users of these components;
- technologies for using these components, allowing the users to meet their subsistence, health, trade and ritual needs; and
- "a view of the world that makes sense of such information, rules, relationships and technologies in the context of a long-term and holistic perspective in decision-making."[35]

2. TK on biological diversity, which is the "knowledge, innovations and practices of indigenous and local communities embodying traditional lifestyles relevant for the conservation and sustainable use of biological diversity."[36] This sub-category of traditional ecological knowledge is "a body of knowledge built by a group of people through generations living in close contact with nature. It includes a system of classification, a set of empirical observations about the local environment, and a system of self-management that governs resource use."[37]
3. Traditional environmental knowledge, which is the knowledge of indigenous peoples that have used the natural resources of their local environments in an ecologically sustainable manner for many generations.[38]
4. Traditional taxonomic knowledge,[39] which is the knowledge on the taxonomy of land, landscape, fauna and flora, and other phenomena of the environment.[40] According to para. 8 of Recommendation II/2 of the Subsidiary Body on Scientific, Technical and Technological Advice (SBSTTA), "[i]t should be recognized that traditional taxonomic systems offer a valuable perspective on biological diversity and should be considered part of the total taxonomic knowledge base at national, regional and subregional levels."[41] And finally

---

[35] Reprinted in *Traditional Related Knowledge and the Convention on Biological Diversity* para. 30, UNEP Doc. UNEP/CBD/SBSTTA/2/Inf.3 (July 19, 1996).

[36] Reprinted in *Traditional Knowledge and Biological Diversity* para. 7, UNEP Doc. UNEP/CBD/TKBD/1/2 (Oct. 18, 1997).

[37] *Id.* para. 84.

[38] Martha Johnson, *Dene Traditional Knowledge*, last visited Dec. 20, 1999, <http://www.carc.org/pubs/v20no1/dene.htm>. According to Johnson, "[o]nly in the past decade has this.... TEK been recognized among the western scientific community for its value to contemporary environmental management. Today, a growing body of literature attests not only to the presence of a vast reservoir of information regarding plant and animal behaviour, but also to the existence of effective systems of self-management of natural resources." (*Id.*).

[39] Regarding traditional taxonomic knowledge, the Institute of Development Studies (IDS) states that "[l]ocal people use many categories in different parts of the world to describe types of land, landscape, crops, wild plant species and other natural resources. The categories and names used by them usually differ from those used by scientists. In addition, the criteria of classification are usually functional, that is, related to use, unlike the standardized categorization criteria derived from physical sciences." (Institute of Dev. Studies [IDS], Univ. of Sussex, *Farmers' Knowledge, Innovations, and Relation to Science*, *in* FARMER FIRST. FARMER INNOVATION AND AGRICULTURAL RESEARCH 31, 31 [Robert Chambers et al. eds., 1989]).
According to Wilson, "[a] few shamans and tribal elders are able to put names on a thousand or more species of plants. Not only do the botanists of Europe and North America generally agree with these species distinctions, but they have learned a great deal from their Amerindian colleagues about the habitat preferences, flowering seasons, and practical uses of the different plants. It is a remarkable fact that the only crop used widely by developed countries not already known to native peoples is the macadamia nut, which originated in Australia." (EDWARD O. WILSON, THE DIVERSITY OF LIFE 43-44 [W. W. NORTON & CO. 1993] [1992]).

[40] According to the Executive Secretary of the Biodiversity Convention, "[i]ndigenous and local community taxonomy systems also extend to non-biological phenomena of the physical environment, e.g., types of ice, soils, seasons, meteorological phenomena, etc." (Reprinted in *Traditional Knowledge and Biological Diversity* para. 74, UNEP Doc. UNEP/CBD/TKBD/1/2 [Oct. 18, 1997]).

[41] Reprinted in *Report of the Second Meeting of the Subsidiary Body on Scientific, Technical and Technological Advice* Recommendation II/2, para. 8, UNEP Doc. UNEP/CBD/COP/3/3 SBSTTA-2 Report (Sept. 7, 1996).

5. traditional resource management knowledge,[42] which is the knowledge on the management of natural resources. Examples of this subcategory of traditional ecological knowledge are burning practices or limitations on the number of animals hunted in a certain geographical area or period of time.

Traditional indigenous knowledge[43]

This form of traditional knowledge is described as a way of life, based on the experience of the individual and of the community, as well as knowledge passed down from one's elders and incorporated in indigenous languages. This knowledge is constantly being adapted to the changing environment of each community and will remain current as long as people still use the land and sea and their resources.[44]

This form of TK is sometimes used as a synonym of traditional ecological knowledge.[45] According to Brush, "[i]ndigenous knowledge is an essential part of the in situ maintenance of crop genetic diversity, and it therefore should be encouraged and valued as much as the other components in the conservation of genetic resources, the germplasm itself and bio-systematics."[46] According to the Executive Secretary of the Biodiversity Convention, indigenous knowledge "[is] likely to be highly specific to the locality which each community inhabits and thus may be of limited application outside of those localities. However, it is because of the innovations which take place under locality-specific conditions that subtle variations in the genetic characteristics of local varieties of particular species occurs as traditional innovators seek to accommodate the local environmental conditions under which they live and nurture useful species. Such conditions include soil types, climatic variations, pests, species composition of the area, etc. It is thus the combination of accumulative knowledge and potential for innovation and adaptation of traditional systems, and the equivalent knowledge base and innovative capacity of modern or scientific systems, which, if encouraged, offers unquantifiable, but probably substantial, opportunities for identifying improved techniques for conservation and sustainable use of biological diversity."[47]

Traditional pharmaceutical knowledge[48]

This form of TK includes traditional healing practices and knowledge on the use of natural resources to be found in the surroundings of indigenous and local communities as pharmaceutical compounds. These communities have, for example, used plants and animals to

---

[42]See *Traditional Knowledge and Biological Diversity* para. 78, UNEP Doc. UNEP/CBD/TKBD/1/2 (Oct. 18, 1997).

[43]The term indigenous can have two meanings: First, it can be used as a synonym for "folk", "local", "nonformal" or "non-Western". (Stephen B. Brush, *Indigenous Knowledge of Biological Resources and Intellectual Property Rights: The Role of Anthropology*, 95 AM. ANTHROPOLOGIST 653, 658, 659 [1993]). Second, the term can be used in a politicized way; in this case, "[i]ndigenous... takes on a more essentialist tone and is glossed with "autochthonous", "native", or "aboriginal". The current use of indigenous is both sociological as descriptive and ideological as proscriptive, suggesting that minority ethnic groups should have more power and recognition." (*Id.* at 659 [emphasis omitted]). See generally *id.* at 658-60.

[44]*Recommendations on the Integration of Two Ways of Knowing: Traditional Indigenous Knowledge and Scientific Knowledge* of the Seminar on the Documentation and Application of Indigenous Knowledge held in Inuvik, Northwest Territories, Canada, on Nov. 15-17, 1996, <http://www.inusiaat.com/tek.htm>, last visited Dec. 20, 1999.

[45]*Id.*

[46]Stephen B. Brush, *Farmers' Rights and Genetic Conservation in Traditional Farming Systems*, 20 WORLD DEV. 1617, 1624-25 (1992).

[47]Reprinted in *Traditional Knowledge and Biological Diversity* paras. 87-88, UNEP Doc. UNEP/CBD/TKBD/1/2 (Oct. 18, 1997).

[48]The company Shaman Botanicals, for example, concentrates its pharmaceutical research and development activities on ethnobotanically used plants. *See generally* <http://www.shamanbotanicals.com> (last visited Feb. 28, 2000).

heal diseases, and have known and valued particular properties of soils such as their anti-tumor, antibiotic, or steroid characteristics.[49]

Traditional agricultural knowledge[50]

This form of TK includes plant breeding techniques and knowledge on particular properties of agricultural plants and animals as well as soil and pest management. A well-known example is the Neem tree that is used in Asian countries as the basis for pesticides.[51]

General characteristics of traditional knowledge

Considering the great diversity of TK existing, finding a definition of this term that applies to all forms of TK proves to be a rather difficult undertaking. Nevertheless, TK generally has some or all of the following characteristics:

1. it is created, improved, used and handed down within indigenous and other local communities;
2. the creation of TK occurs over several generations, with each generation adding to and improving the existing TK;
3. the creation and improvement of TK is often a group effort of some or all members of an indigenous or local community; TK is thus the result of a cooperative[52] or informal[53] innovation system;
4. TK is closely connected to the way of life of its users and the environment they live in;[54]
5. TK often concerns ecological aspects such as the sustainable use and management of natural resources;
6. TK is created to support and improve the livelihood of its users and to enable their survival, not to make profits;

[49]RURAL ADVANCEMENT FOUND. INT'L (RAFI), CONSERVING INDIGENOUS KNOWLEDGE: INTEGRATING TWO SYSTEMS OF INNOVATION 23 (1994). For examples of such properties see *id.*

[50]See generally, e.g., MARTIN A. GIRSBERGER, BIODIVERSITY AND THE CONCEPT OF FARMERS' RIGHTS IN INTERNATIONAL LAW: FACTUAL BACKGROUND AND LEGAL ANALYSIS (Thomas Cottier ed., Series on Global Economic Law No. 1, 1999).

[51]The following are examples of the use of landraces and TK regarding landraces in modern innovations:

- Sally V. Fox, an American seed company, increased through conventional plant breeding techniques the staple length of traditional colored Andean cotton to enable its commercial weaving, and was granted plant breeders' rights for this colored cotton. (RURAL ADVANCEMENT FOUND. INT'L (RAFI), CONSERVING INDIGENOUS KNOWLEDGE: INTEGRATING TWO SYSTEMS OF INNOVATION 9 (1994).
- The British Agricultural Genetics Company was awarded a patent for the Cowpea-Trypsin Inhibitor (cpTI) gene. It is assumed that the used genetic material originated in Africa. The sales value of this cpTI gene is estimated to be several hundred million USD. (*Id.* at 20). RAFI states that "[w]hile indigenous farmers in Africa might not have been aware of the specific cpTI gene, they were aware of the utility of the plant in inhibiting insect pests." (*Id.*).

[52]"Indigenous knowledge has not been the result of passive accumulation. On the contrary, it is the result of a dynamic "cooperative innovation system" that continues to work and continues to offer humankind hope for planetary survival." (Sarah L. Timpson & Luis Gomez-Echeverri, *Foreword* to RURAL ADVANCEMENT FOUND. INT'L [RAFI], CONSERVING INDIGENOUS KNOWLEDGE: INTEGRATING TWO SYSTEMS OF INNOVATION at i, ii [1994]).

[53]According to RAFI, the results of formal and informal innovation systems differ as follows: "[a]s a fair simplification, it can be said that the institutional system offers humanity micro-system developments that find application on a macro-scale. Highly specific improvements in molecular biology or micro-electronics may have vast commercial application. The cooperative system, on the other hand, offers broad macro-system innovations that generally can only be applied at the micro-level, i.e., the local environment. Indigenous knowledge often involves the use of complex bio-systems integrating plants, insects, and soil, for example, in a common strategy." (RURAL ADVANCEMENT FOUND. INT'L [RAFI], CONSERVING INDIGENOUS KNOWLEDGE: INTEGRATING TWO SYSTEMS OF INNOVATION 2 [1994]).

[54]The Executive Secretary of the Biodiversity Convention states that "much traditional knowledge will mean little outside the environment where it arises and is likely to be most valuable as a means to achieve *in-situ* ecosystem management." (Reprinted in *Traditional Knowledge and Biological Diversity* para. 89, UNEP Doc. UNEP/CBD/TKBD/1/2 [Oct. 18, 1997]).

7. access to and use of TK within an indigenous or local community is regulated by local laws, customs and traditions; even though the ownership of TK may vary, it is usually beneficial to all members of the indigenous or local community; and
8. outside the scope of these local laws, customs and traditions, TK belongs to the public domain. With these characteristics, TK can be contrasted with "modern", "scientific",[55] "Western" or "cosmopolitan"[56] knowledge.

The protection of traditional knowledge by intellectual property rights

Aims of the protection of traditional knowledge

TK has been protected by local laws, customs and traditions for many centuries. These local rules have not only allowed for the development of the enormous wealth of TK in existence today, but they have also regulated access to and use of TK within indigenous and local communities and its handing down to the next generation. Over the years, these local rules have not been static, but have been adapted to changing circumstances and needs instead.

If these local rules have so far been adequate, the question arises why new forms of protection for TK are necessary. Two main reasons can be advanced for the necessity of new rules: First, in contrast to past times, TK is increasingly used outside the range of application of these local rules. Some Western scientists, for example, are using the traditional pharmaceutical knowledge of local and indigenous communities in modern pharmaceuticals. The use of TK by entities other than the indigenous and local communities renders these local rules inadequate and necessitates new rules that regulate access to and use of TK by these other entities. Second, often caused by factors that lay beyond the sphere of influence of its holders, TK is not developed, used and handed down as it was in the past. As TK continues to be important for a large part of the world's population, this loss affects the daily life of millions of people and should thus be halted or, if possible, even be reversed.

Depending on what is aimed at with the realization of new forms of protection of TK, different systems of protection will be in the foreground. The first and most basic issue that should thus be clarified is the aim of the protection of TK. Without clarifying this issue, the realization of the protection of TK carries the risk of being ill-adapted in practice and may thus even have counterproductive effects. In literature and in discussions in international fora, proposals for the protection of TK often seem to be advanced without considering this issue. Even though these proposals are undoubtedly made with the best of intentions, they thus do not address one of the key issues of the protection of TK.

The protection of TK can have different aims, including the following: it can aim at

1. granting the holders of TK the right to have a voice in the decision making process regarding access to, use and commercialization of TK by entities other that its original holders;
2. preventing the unjustified appropriation of TK by entities other than its original holders;[57]

---

[55]See, e.g., *Recommendations on the Integration of Two Ways of Knowing: Traditional Indigenous Knowledge and Scientific Knowledge* of the Seminar on the Documentation and Application of Indigenous Knowledge held in Inuvik, Northwest Territories, Canada, on Nov. 15-17, 1996, <http://www.inusiaat.com/tek.htm>, last visited Dec. 20, 1999.

[56]The Secretariat of the Biodiversity Convention states that TK "can be contrasted with cosmopolitan knowledge, which is drawn from global experience and combines "western" scientific discoveries, economic preferences and philosophies with those of other widespread cultures." (Reprinted in *Traditional Related Knowledge and the Convention on Biological Diversity* para. 22, UNEP Doc. UNEP/CBD/SBSTTA/2/Inf.3 [July 19, 1996]).

[57]The Recommendations adopted at the Seminar on the Documentation and Application of Indigenous Knowledge state that TEK should be documented and, with community approval, made readily accessible to the scientific community so that it may be referenced and properly cited. In this way, the holders of TEK can be given full recognition and the information will not be regarded as anecdotal data or as a "new discovery" when scientists use it. (*Recommendations on the Integration of Two Ways of Knowing: Traditional Indigenous Knowledge and Scientific Knowledge* of the Seminar on the Documentation and Application of Indigenous Knowledge held in Inuvik, Northwest Territories, Canada, on Nov. 15-17, 1996, <http://www.inusiaat.com/tek.htm>, last visited Dec. 20, 1999).

3. creating a new form of property rights, which would act as the counterpart of existing forms of IPRs;
4. furthering the development, sustainable use and conservation of TK;[58] and finally
5. combining all of the above mentioned aims.

Systems available to protect traditional knowledge

Several systems are available to protect TK,[59] which can be categorized according to different factors. First, TK can be protected by legal or non-legal means of protection. Examples of legal means are property rights, whereas incentive measures and capacity-building measures such as the transfer of technology are examples of non-legal means. Second, TK can be protected by contractual arrangements or statutory provisions.[60] Contractual arrangements such as material transfer agreements, non-disclosure agreements or information transfer agreements[61] would regulate access to and use of TK on a bilateral level, that is, between the entity with the right of disposal over the TK and the entity interested in accessing and using this TK. Statutory provisions would regulate access to and use of TK in a general way, that is, they would apply in all cases where TK is accessed and used. Statutory provisions can be realized on the national or international level, and they can be of legally binding or legally non-binding nature. Third, TK can be protected by *sui generis* forms of protection, that is, protection that is specifically designed for the task of protecting TK. This can be *sui generis* forms of IPRs or other *sui generis* systems of protection. Fourth, TK can be protected by exclusive or non-exclusive rights,

---

[58]According to the Executive Secretary of the Biodiversity Convention, the major thrust for the development of appropriate incentives should be to secure the survival of traditional knowledge within and beyond this generation, that is, such incentives should also be targeted towards the next generations to ensure that it is worth their while to carry the knowledge and skills of their forebears....It should also be borne in mind that the conservation of the knowledge, skill, practice and spirit of innovation cannot take place without looking at the socio-cultural and ecological context of the knowledge system.... It should be noted that no one incentive will be appropriate for all situations within a community or across communities....Most importantly, incentive measures should be developed and applied in such a way as to maintain community and ecological balance. (Reprinted in *Traditional Knowledge and Biological Diversity,* paras. 61-63, UNEP Doc. UNEP/CBD/TKBD/1/2 [Oct. 18, 1997]).
The participants in the Seminar on the Documentation and Application of Indigenous Knowledge state that "[i]n all circumpolar regions represented, the elders who grew up living on the land and sea are passing away, and with them their knowledge of the land and of the traditions of the community. It is essential both that we document the elders' knowledge, and that we promote the continued transmission of that knowledge to today's youth, which must include protecting the use of the land and its resources. Otherwise, traditions and knowledge will be lost, and the people who remain and the generations to follow will be left without the benefit of this irreplaceable heritage." (*Recommendations on the Integration of Two Ways of Knowing: Traditional Indigenous Knowledge and Scientific Knowledge* of the Seminar on the Documentation and Application of Indigenous Knowledge held in Inuvik, Northwest Territories, Canada, on Nov. 15-17, 1996, <http://www.inusiaat.com/tek.htm>, last visited Dec. 20, 1999). See *id.* for the recommendations made by indigenous peoples of the Arctic region regarding the perpetuation, documentation, application and use of TK.
[59]See generally, e.g., *Legal and Other Appropriate Forms of Protection for the Knowledge, Innovations and Practices of Indigenous and Local Communities Embodying Traditional Lifestyles Relevant for the Conservation and Sustainable Use of Biological Diversity*, UNEP Doc. UNEP/CBD/WG8J/1/2 (Jan. 10, 2000).
[60]The Executive Secretary of the Biodiversity Convention states that "[r]eliance upon such contractual methods to capture benefit for indigenous and local communities is widely thought of as the most practical approach to ensure the equitable sharing of benefits referred to in Article 8(j)." (Reprinted in *Traditional Knowledge and Biological Diversity* para. 41, UNEP Doc. UNEP/CBD/TKBD/1/2 [Oct. 18, 1997]). *See generally, e.g.,* Steven M. Rubin & Stanwood C. Fish, *Biodiversity Prospecting: Using Innovative Contractual Provisions to Foster Ethnobotanical Knowledge, Technology, and Conservation*, 5 COLO. J. INT'L ENVTL. L & POL'Y 23 (1994).
[61]Reprinted in *Traditional Knowledge and Biological Diversity* para. 40, UNEP Doc. UNEP/CBD/TKBD/1/2 (Oct. 18, 1997).

the former consisting primarily of existing or new forms of IPRs. And finally fifth, TK can be protected by rights held by individuals or group rights.

The following analysis focuses on existing and new forms of IPRs to provide the protection of TK.

The suitability of intellectual property rights to protect traditional knowledge

When determining the suitability of existing or new forms of IPRs to protect TK, the following points need to be considered:

To be protectable by existing or new forms of IPRs, TK must meet the applicable criteria of protection. Examples of such criteria of protection are novelty, industrial applicability, stability and homogeneity. All TK not meeting these criteria of protection is left without protection and thus remains in the public domain.

To be able to apply for and enforce IPRs, the holders of TK need access to technical, legal and financial means. Indigenous and local communities often lack these means. Furthermore, widespread illiteracy and the use of languages and dialects that often differ substantially from the official language additionally complicate these activities. In this respect, NGOs could play an important role and assist indigenous and local communities in applying for and enforcing IPRs protecting TK.

Of the existing forms of IPRs, patents, plant breeders' rights, trade secrets, trademarks and geographical indications are in the foreground to protect TK. Whereas patents and trade secrets allow for the direct protection of TK, plant breeders' rights, trademarks and geographical indications protect the goods incorporating TK and thus only indirectly protect TK.

IPRs are designed to function in a market economy. The creation and further development of TK, however, generally takes place outside a market economy.

The protection of TK by IPRs could thus disrupt the traditional way of creating and developing TK.

With the increasing globalization, non-market-based economies are likely to come under further pressure. Protecting TK by IPRs would allow indigenous and local communities to participate in the global market-based economy.

The value of IPRs is determined by the market for the subject matter protected by these IPRs. If the demand for this subject matter is high, the value of the IPRs is also high; in contrast, if the demand low, the value is also low. At present, however, no real market exists for TK; thus, if IPRs are used to protect TK, their value is likely to be low.

The returns created by IPRs protecting TK are likely to be small. It seems rather questionable whether these returns will be sufficient to cover all expenses involved in protecting TK by IPRs. If so, protecting TK by IPRs would be senseless from an economic point of view. Furthermore, if the protection of TK aims at the conservation of this form of knowledge, adequate financial means must be available. If, however, the returns created by IPRs are the main source of these financial means, conservation efforts will be greatly limited due to the lack of financial means.

IPRs are a form of private property rights. Some indigenous and local communities, however, reject the notion of private property for religious or ideological reasons.[62] Protecting TK exclusively by IPRs would force these communities either to accept the notion of private property against their beliefs or to forego this protection. Clinton states that Indians and many non-westerners, often have a very different view of the nature of their rights and legal relationships. Deriving their legal vision from their tribal associations, tribal traditions, and the

[62]According to RAFI, most indigenous peoples share a sense of communal responsibility for their land and its living resources. It is rare to find a deeply-rooted culture that permits a patent-like monopoly over the products or processes of life. It is largely because of this communal tradition that many indigenous peoples look upon intellectual property – especially related to life forms – as a kind of blasphemy. (RURAL ADVANCEMENT FOUND. INT'L [RAFI], CONSERVING INDIGENOUS KNOWLEDGE: INTEGRATING TWO SYSTEMS OF INNOVATION 3 [1994]).

natural ecology with which they often seem more familiar than many western political philosophers, native peoples see humans as inherently social beings. As social beings, people never exist isolated from others in some mythic, disorganized state of nature. Rather human beings are born into a closely linked and integrated network of family, kinship, social and political relations. One's clan, kinship and family identities are part of one's personal identity and one's rights and responsibilities exist only *within* the framework of such familial, social, and tribal networks. Non-Western thinkers, therefore, naturally think of their rights as part of a group. Certain rights exist within each social group and other rights and responsibilities are attendant to their relations with members of other groups within the web of associations that forms the tribe or the state.[63]

IPRs, as exclusive rights, can only be awarded to specific holders, that is, specific natural or legal persons, for specific intangibles. In the content of TK, determining not only the specific holders, but also specific intangibles will often prove to be very difficult, if not impossible.

Governments need to implement adequate administrative procedures for the application and/or registration of IPRs protecting TK. The same can be said about the enforcement of IPRs, which requires a well functioning judicial system and appropriate legal remedies. With the entering into force of the TRIPS Agreement on January 1, 2000 for most developing countries, minimal standards for the administration and enforcement of IPRs should be in place in the majority of the Member States of the TRIPS Agreement.

TK is generally held by indigenous and local communities.[64] These holders often belong to the political, religious, ethical or economic minorities. Their minority status will often prevent them from having access to the necessary financial, technical and legal means. It thus seems questionable whether the holders of TK will be able to apply for, benefit from and enforce their IPRs.

Conclusion

The protection of TK has not only been an issue in literature, but also the subject of debate in international fora such as the Biodiversity Convention, WIPO and the WTO. Some of these debates have been very controversial and have not yet produced concrete results. Considering the many legal, scientific, economic and political questions arising when realizing this protection, the lack of any concrete results comes as no surprise.

This paper provides a summary of the background of the problematic area of IPRs and TK. This summary makes apparent that a great deal of issues need to be considered when discussing any proposals for the protection of TK. Considering all issues concerned, however, is crucial for developing viable answers to the many questions arising. Otherwise, the protection of TK realized runs the risk of being ill-adapted in practice and may even have counterproductive effects.

Clarifying the relevant terminology is also crucial for the debate on IPRs and TK. A more detailed analysis of documents and statements made at international conferences reveals that the term IPRs is sometimes used in a broader sense than the specific meaning this term has in law.

---

63 Robert N. Clinton, *The Rights of Indigenous Peoples as Collective Group Rights*, 32 ARIZ. L. REV. 739, 742 (1990).

64 According to Brush "[g]roups that control the greatest wealth of indigenous knowledge and biological resources may be the most marginalized by nation-states – for example the Kurds of Iraq or the Mayan peasants of Guatemala. Few states accept the rights of indigenous groups *per se*, and even fewer can be expected to lend their political and legal apparatus to promote intellectual property rights to indigenous groups.... The history of prejudice and discrimination against indigenous peoples and ethnic minorities around the world suggests that the benefits from intellectual property protection for biological resources and indigenous knowledge will not reach indigenous people." (Stephen B. Brush, *Indigenous Knowledge of Biological Resources and Intellectual Property Rights: The Role of Anthropology*, 95 AM. ANTHROPOLOGIST 653, 664-65 [1993]).

To avoid unnecessary misunderstandings, this paper suggests to use alternative terms such as "intellectual achievements", "intellectual contributions" or "intellectual heritage" whenever the term IPRs is used in this broader sense. The description of TK has shown that a great deal of differing forms of TK exist, preventing the creation of a generally applicable definition of this term. It has also shown that the terminology used in literature is not always uniform, further complicating the issue. Nevertheless, several characteristics can be found that generally apply to all forms of TK.

In literature and statements made at international conferences, proposals for the protection of TK often seem to be made without first clarifying the aims of the proposed protection. This clarification, however, is necessary, as the aim of the protection of TK will be decisive for the choice of the system of protection. The following two aims are frequently mentioned in the debate on the protection of TK and are likely to be in the foreground: First, the increasing use of TK by entities from outside the indigenous and local communities that hold the TK in question has rendered the local rules regulating access to, use and handing down of TK inadequate. New rules are thus necessary that consider the particular circumstances. And second, due to various factors that prevent the development, use and handing down of TK as in past times, TK is lost at increasing rates.

Different systems have been proposed to protect TK. One system frequently advanced is that of IPRs. This paper analyses the suitability of IPRs to protect TK and discusses the arguments speaking in favor of and against this system of protection. This analysis shows that IPRs are inadequate to provide the comprehensive protection of TK, especially if they are the sole system of protection.

Rather, TK should be protected by a new form of socio-economic right. This socio-economic right should allow to meet both aims, that is, it should not only regulate access to and use of TK by entities from outside the indigenous and local communities, but also allow for the conservation and sustainable use of TK. This can only be achieved with a bundle of rights and measures. Viable elements could be:

1. guidelines regulating access to and use of TK and the sharing of the benefits arising from this use;
2. national access legislation;
3. where suitable, the application of existing forms of IPRs or the realization of new forms of IPRs;
4. government policies protecting the natural surroundings of the indigenous and local communities and their way of life;
5. a register of TK operated by an international organization such as WIPO;
6. measures enabling indigenous and local communities to better manage their own affairs;
7. the transfer of technical, legal and scientific know-how; and finally
8. adequate funding.

The author is aware that the proposed new socio-economic right remains rather vague and poses many difficult to answer legal, scientific, economic and political questions. Furthermore, some of the proposed elements of the new socio-economic right are, from a political point of view, very controversial. Considering the different interests involved in the debate on the protection of TK, the international community is still a long way from adequately solving this issue.

References

Agreement on Trade-Related Aspects of Intellectual Property Rights (1994) adopted Dec. 15, 1993, reprinted in 33 I.L.M. 1197

Baer LA., Initiatives for the Protection of Rights of Holders of Traditional Knowledge, Indigenous Peoples and Local Communities, document prepared for the WIPO Roundtable on Intellectual Property and Indigenous

Peoples, Geneva, July 23-24, 1998, last visited Dec. 20, 1999 <http://www.wipo.int/eng/meetings/1998/indip/rt98_4a.htm>

Breckenridge LP (1992) Protection of Biological and Cultural Diversity: Emerging Recognition of Local Community Rights in Ecosystems Under International Environmental Law, 59 *Tenn. L. Rev. 735*

Brush SB (1992) Farmers' Rights and Genetic Conservation in Traditional Farming Systems, 20 *World Dev.* 1617

Brush SB (1993) Indigenous Knowledge of Biological Resources and Intellectual Property Rights: The Role of Anthropology, 95 *Am. Anthropologist* 653

Chambers R, Pacey A, Thrupp LA (eds.) (1989) Farmer First. *Farmer Innovation and Agricultural Research* (Intermediate Technology Publications, London)

Clinton RN (1990) The Rights of Indigenous Peoples as Collective Group Rights. 32 *Ariz. L. Rev.* 739

Convention on Biological Diversity (1992) adopted June 5, 1992, reprinted in 31 I.L.M. 818

Cottier T (1995) The Protection of Intellectual Property Rights: A Requirement for Technology Cooperation, Foreign Investment and Equitable Returns in Biotechnology Prospecting. In: Schweizerisches Zentrum für Internationale Landwirtschaft [ZIL] (ed.): *Biotechnologie für Entwicklungsländer? Chancen und Risiken der Biotechnologie bei landwirtschaftlichen Nutzpflanzen* 65

Girsberger M (1999) Biodiversity and the concept of farmers' rights in international law: factual background and legal analysis. *Series on Global Economic Law* No. 1

Girsberger M (1998) The Protection of Traditional Plant Genetic Resources for Food and Agriculture and the Related Know-How by Intellectual Property Rights in International Law – The Current Legal Environment, 1 *J. World Intell. Prop.* 1017

Glowka L, Burhenne-Guilmin F, Synge H (1994) A guide to the Convention on Biological Diversity (The World Conservation Union (IUCN), Gland, Switzerland and Cambridge, UK

Hurlbut D (1994) Fixing the Biodiversity Convention: Toward a Special Protocol for Related Intellectual Property. 34 *Nat. Resources J.* 379

Johnson M (1999) *Dene Traditional Knowledge*, last visited Dec. 20, 1999 <http://www.carc.org/pubs/v20no1/dene.htm>

Legal and Other Appropriate Forms of Protection for the Knowledge, Innovations and Practices of Indigenous and Local Communities Embodying Traditional Lifestyles Relevant for the Conservation and Sustainable Use of Biological Diversity, UNEP Doc. UNEP/CBD/WG8J/1/2 (Jan. 10, 2000)

Odek JO (1994) Bio-Piracy: Creating Proprietary Rights in Plant Genetic Resources. 2 *J. Intell. Prop. L.* 141.

Recommendations on the Integration of Two Ways of Knowing: Traditional Indigenous Knowledge and Scientific Knowledge of the Seminar on the Documentation and Application of Indigenous Knowledge held in Inuvik, Northwest Territories, Canada, on Nov. 15-17, 1996, last visited Dec. 20, 1999, <http://www.inusiaat.com/tek.htm>

Report of the Panel of Experts on Access and Benefit-Sharing UNEP Doc. UNEP/CBD/COP/5/8 (Nov. 2, 1999)

Report of the Second Meeting of the Subsidiary Body on Scientific, Technical and Technological Advice Recommendation II/2, UNEP Doc. UNEP/CBD/COP/3/3 SBSTTA-2 Report (Sept. 7, 1996)

Report of the Workshop on Traditional Knowledge and Biological Diversity, UNEP Doc. UNEP/CBD/TKBD/1/3 (Dec. 15, 1997)

Ritchie M, Dawkins K, Vallianatos M: Intellectual Property Rights and Biodiversity: The Industrialization of Natural Resources and Traditional Knowledge, 11. *St. John's J. Legal Comment.* 431 (1996)

Roht-Arriaza N (1996) Of Seeds and Shamans: The Appropriation of the Scientific and Technical Knowledge of Indigenous and Local Communities. 17 *Mich. J. Int'l L.* 919.

Rubin SM, Fish SC (1994) Biodiversity Prospecting: Using Innovative Contractual Provisions to Foster Ethnobotanical Knowledge, Technology, and Conservation. 5 *Colo. J. Int'l Envtl. L & Pol'y* 23 (1994)

Rural Advancement Found Int'l (RAFI) (1994) Conserving indigenous knowledge. Integrating two systems of innovation

Serote MW (1998) Initiatives for Protection of Rights of Holders of Traditional Knowledge, Indigenous Peoples and Local Communities. document prepared for the WIPO Roundtable on Intellectual Property and Indigenous Peoples, Geneva, July 23-24, 1998, last visited Dec. 20, 1999 http://www.wipo.int/eng/meetings/1998/indip/rt98_4c.htm

The Crucible Group (1994) People, Plants and Patents: The Impact of Intellectual Property on Trade, Plant Biodiversity and Rural Society (International Development Research Centre, Ottawa, Canada)

Traditional Knowledge and Biological Diversity UNEP Doc. UNEP/CBD/TKBD/1/2 (Oct. 18, 1997)

Traditional Related Knowledge and the Convention on Biological Diversity, UNEP Doc. UNEP/CBD/SBSTTA/2/Inf.3 (July 19, 1996)

Wilson EO (1993) The Diversity of Life (WW Norton & Co., New York)

Woodliffe J (1996) Biodiversity and Indigenous Peoples. In: Michael Bowman and Catherine Redgwell (eds.): International Law and the Conservation of Biological Diversity, 225-269 (Kluwer Law International, London)

© 2003 Birkhäuser Verlag Basel/Switzerland

The Cartagena Protocol on Biosafety of the Convention on Biological Diversity: overview and future development

Beat Nobs, Robert Lamb and François Pythoud

Swiss Agency for Environment, Forests and Landscape, CH-3003 Bern, Switzerland

Overview of scope and objectives of the convention on biological diversity

The Convention on Biological Diversity (CBD)[1] was negotiated under the United Nation Environment Programme (UNEP). The final text was adopted in 1992 and opened for signature at the United Nations Conference on Environment and Development held in Rio de Janeiro the same year. The Convention came into force on the 29th of December 1993. Today 177 countries including all industrialised countries with the exception of the USA have ratified and are Parties to the Convention.

The three objectives of the Convention are the conservation of biological diversity, the sustainable use of its components and the fair and equitable sharing of the benefits arising from the use of genetic resources. This balance between conservation of biodiversity , its sustainable use and the sharing of benefits is the heart of the political agreement upon which the CBD is founded and articulated. The contracting Parties have the responsibility to take all relevant measures to achieve these three objectives. Even though most of these measures involve actions at national level, some require intensive international collaboration. This is specially the case for issues associated with sustainable use of genetic resources including biotechnology and benefit sharing.

Biotechnology and biosafety in the convention on biological diversity

Biotechnology has been closely associated to the development of the CBD since the beginning. The Convention defines biotechnology as "any technological application that uses biological systems, living organisms, or derivatives thereof to make or modify products or processes for specific uses". Issues related to biotechnology fall into four categories: safety, access to genetic resources, technology transfer and benefit sharing.

Access to genetic resources versus access to biotechnology: the deal behind the Convention Biotechnology relies primarily on biological and genetic resources, which are largely located in the countries from the South. In order to allow those countries to fulfil their obligation to protect these resources and to facilitate their access, the Convention foresees specific measures in the area of benefit sharing and transfer of technology. Article 15 of the Convention requires the establishment of mechanisms to insure fair and equitable sharing of the benefits arising from the commercial and other use of these genetic resources. Article 16 requests the Parties to facilitate the access to and transfer of technologies including expressly biotechnology. In this process, Parties should take appropriate measures to promote the participation of the private sector and

[1]The full text of the Convention and all related informations are available at http://www.biodiv.org.

to ensure that countries supplying genetic resources, especially the developing countries, are provided access to all technologies and in particular biotechnology that make use of these genetic resources. In the same way, and although somehow redundant article 19 reemphasises again the importance of biotechnology and the distribution of the benefits arising from its use. In particular, each Party must promote priority access to the results and benefits arising from biotechnology based upon genetic resources to country supplying the genetic resources. This clearly shows the importance of the transfer of biotechnology in this political agreement aiming to respond to the strong demand of developing countries to share the benefit of the use of the genetic resources that they have to protect and conserve.

Biosafety: a condition for the sound transfer of biotechnology

In the beginning of the 1990s, almost all industrialised countries were developing domestic environmental regulations to control the experimental and commercial use of genetically modified organisms (GMO). Due to this increasing regulatory burden in industrialised countries, it was feared that some of the large multinational companies might pursue testing of transgenic organisms in developing countries where no regulations were in place. This evolution and context explain why, during the final stage of negotiation of the Convention in 1992, there was a strong pressure from almost all developing countries, to integrate into the Convention internationally legally binding measures to regulate the transfer, handling and use of living modified organisms (LMO). At that time, mainly due to the strong opposition of some OECD countries with an active biotechnology sector, the decision to embark on a negotiation for a Protocol on biosafety was deferred in article 19.3 of the CBD to the Conference of the Party (CoP). In article 19.3 of the CBD, Parties to the Convention are invited "...to consider the need for and modalities of a protocol setting out appropriate procedures, including in particular, advance informed agreement in the field of safe transfer, handling and use of any living modified organisms resulting from biotechnology that may have adverse effects on the conservation and sustainable use of biological diversity."

As a result of this whole negotiation, the only concrete obligations addressing biosafety under the CBD were the provisions of article 8 g which require Parties to establish and maintain domestic regulatory framework to address the environmental and health effects associated with the use of biotechnology and the provisions of Article 19.4 regarding international co-operation and exchange of information. This was obviously not sufficient to insure the safe transfer and trade of biotechnology and living modified organisms from the North to the South.

A brief history of the negotiation of the biosafety protocol

In 1995 in Jakarta the 2nd Conference of the Parties to the CBD came to the conclusion that there was a need for an international legal instrument to address the issue of biosafety. Therefore COP2 decided to launch a negotiation process to develop a Protocol on biosafety specifically focusing on the transboundary movement of LMOs. For this purpose, the Biosafety open-ended Ad Hoc Working Group (BSWG) was established. Between 1996 and 1999, the BSWG met six times. In 1998, noting the progress made during the negotiation, the 4th Conference of the Parties to the CBD decided that the protocol would be adopted by the 1st extraordinary Conference of the Parties (ExCoP) in Cartagena (Colombia) in February 1999. Unfortunately the Cartagena Conference was unsuccessful. No agreement was found and the ExCoP was suspended. The main remaining divergences were the following:

- The transboundary movement of LMOs intended for use as food, feed or for processing (LMOs-FFP) in other word LMOs that are traded as agricultural commodities. The so-called

Miami Group composed of the large agricultural exporting countries wanted to restrict the scope of the AIA procedure under the Protocol to the transboundary movement of LMOs, which were intended to be intentionally introduced in the environment, such as seeds. On the contrary, others were arguing for a broader scope that would include all LMOs including the one which were agricultural commodities. The question of identification and documentation was also clearly associated with this issue.

- The relationship between the Protocol and other international agreements, in particular the WTO. Countries of the Miami Group feared that the trade related provisions of the WTO could be challenged by some provisions of the Protocol, thus creating trade barriers and therefore wanted to introduce a "saving clause" in the protocol. If adopted, such a saving clause would have meant that, in the case of dispute settlement, the existing WTO rules would have prevailed on the Protocol and undermined its effectiveness. Indeed, this would have created a hierarchy between the WTO and the Protocol and moreover, it could have weakened other environmental agreements. This issue emphasise the wider and important political implications of the Protocol which go beyond the limited scope of biosafety. It affects the relation that needs to be clarified between the Trade Agreements (WTO) and the multilateral system of Environment Agreements (MEA).
- The application of the precautionary approach in the context of the Protocol. The precautionary approach implies in essence that preventive action should be taken, even without full scientific certainty. Some feared that the precautionary principle could be an excuse to restrict trade to protect domestic producers. They argued that such restrictions had to be based only on sound science and rigorous risk assessment. Others argued that the sound-science argument itself was an excuse to limit the use of an established principle of international environmental law.

After one year of intensive informal consultation, the 1st extraordinary Conference of the Parties to the CBD reconvened in Montreal from the 24 to the 28 of January 2000 under the Chairmanship of Mr. Juan Mayr, Minister of the Environment of Columbia. After several days and nights of intensive negotiation, a compromise on the contentious issue was finally found and the Protocol was adopted by the ExCoP in the early hours of 29 January 2000 as the Cartagena Protocol on biosafety in recognition of the critical role played by the government of Colombia in the success of the negotiation.

The Cartagena Protocol on Biosafety[2]

The Objective of the Cartagena Protocol is to contribute, in accordance with the precautionary approach, to ensuring an adequate level of protection in the field of the safe transfer, handling and use of LMOs that may have adverse effects on the conservation and sustainable use of biological diversity, taking also into account risks to human health by specifically focusing on transboundary movement. To achieve this objective the Protocol establishes an advanced informed agreement procedure (AIA), incorporates the precautionary approach in the decision process and details information and documentation requirements. The Protocol also contains provisions regarding documentation, information-sharing, capacity building and financial resources with special attention to the situation of developing countries and those without domestic regulatory framework on biosafety. Last but not least the Protocol also clarifies the relation with other international instruments, addresses the socio-economic considerations and set up a process to embark upon further international work on liability and redress.

[2]The text of the Cartagena Protocol is available at http://www.biodiv.org/biosafe/protocol/Protocol.html.

The advance informed agreement procedure (AIA)

The AIA procedure is the backbone of the Protocol. It is addressed mainly in articles 7 to 10 and 12 of the Protocol. The AIA shall apply prior to the first transboundary movement of a LMO intended for intentional introduction into the environment. The Party of export shall take measures to ensure the notification of the information specified in Annex I of the Protocol to the competent national authority of the Party of import. Then Party of import must acknowledge the receipt of notification within 90 days and decide whether to proceed with the AIA decision's procedure, or according to its domestic legislation if this legislation is consistent with the Protocol. The decision procedure under the AIA is driven by the following considerations:

- the decision must be based on a risk assessment carried out in a scientifically sound manner in accordance with Annex II of the Protocol;
- the decision can be implicit or explicit. In case of explicit decision, the Party of import shall communicate its decision within 270 days of the date of receipt of notification. The Party of import could extend this period by a defined period of time after having informed the notifier;
- the decision shall set out the reasons on which it is based except in case of unconditional consent;
- the lack of scientific certainty due to insufficient relevant scientific information regarding the extent of potential adverse effects of a LMO shall not prevent the Party of import to take a decision, as appropriate in order to avoid or minimise such potential adverse effects.
- No decision within 270 days shall not imply consent to the transboundary movement of the LMO.
- The Party of Import could review decision on its own or under request by the Party of export or the notifier in light of change in circumstances or new scientific and technical information.

The special procedure for LMOs intended for use as food, feed or processing (LMO-FFP) agricultural commodities

Under the Protocol, the transboundary movement of LMO-FFP is subject to a special procedure, which could be described as "Advanced information Procedure". A Party that takes a final decision regarding the domestic use of an LMO that may be subject to enter into the international trade of LMO-FFP must inform the other Parties by providing the information specified in Annex II of the Protocol to the Biosafety Clearing House. While the Exporting party basically drove the AIA procedure, the Importing Party mainly drives the procedure for LMO-FFP commodities. The Protocol recognises the right of each Party to take a decision regarding the first import of an LMO-FFP under its domestic regulatory framework if this framework is consistent with the objective of the Protocol. More interesting the Protocol foresees the possibility for developing country which do not have domestic regulatory framework to deal with LMO-FFP to declare that they might take decision regarding the first import of a LMO-FFP in a predictable time frame on the basis of a risk assessment undertaken in accordance with Annex II of the Protocol. This subtle balance between the interest of exporters of agricultural commodities and the needs of developing countries lacking domestic regulatory framework was certainly the key for the success of the last round of negotiation.

Identification and documentation requirements under the protocol

The question of identification in documentation accompanying the transboundary movement of LMOs was the last issue to be resolved in the negotiation. All delegations agreed on the need for clear identification in the case of transboundary movement of LMOs intended for contained use or intentional introduction in the environment. In the case of LMOs intended for intentional introduction, the Protocol even requires additional specification of the identity and the relevant traits and characteristics of the LMOs. On the opposite the discussion on the identification requirements for LMO-FFP was more controversial and not fully settled in the Protocol. The introduction of the clear identification of shipment containing LMO-FFP as LMOS would have introduced a de facto requirement for segregation between LMOs and non-LMOs. The main exporters of agricultural commodities were strongly oppose to this approach mainly because they said they would need more time to put in place a workable system. As a consequence the Protocol requires that shipment-containing LMO-FFP shall be identified as "may contain LMOs" and as "not intended for intentional introduction into the environment". Moreover the Protocol states that the Meeting of the Parties shall, in two years after the entry into force of the Protocol, take a decision on the detailed requirements for documentation of LMO-FFPs including specification of their identity and any unique identification.

Application of the precautionary approach

The Protocol is based on the precautionary approach contained in Principle 15 of the Rio Declaration on Environment and Development as it is clearly stressed both in the Preamble and in the objective. The Protocol goes even one step further in the application of the precautionary approach by clearly stating in Article 10 and 11 that decisions taken by the Party of Import in order to avoid or minimise potential adverse effects of LMOs shall not be prevented by the lack of scientific certainty due to insufficient relevant scientific information. This mainly differs from the approach already existing in the Sanitary and Phtyosanitary Agreement of WTO by the fact that the decision is not necessarily provisional and that it could also be based on potential adverse ecological impacts.

The relation with other international agreement

This contentious political issue is resolved by the formulation contained in the Preamble of the Protocol that recognises that international agreements are of equal status and that trade and environment agreements should be mutually supportive with a view to achieving sustainable development. This concept of mutually supportiveness could serve in the future as a model to clarify the relation between the WTO and other Environmental International Agreements.

The Biosafety Clearing House: the central tool for the implementation of the protocol

The implementation of the Protocol requires access to and exchange of a broad range of scientific, technical environmental and legal information on LMOs and on biosafety. To fulfil those needs, the Protocol establishes the Biosafety Clearing House as a part of the Clearing House Mechanism of the Convention on biological diversity. The Biosafety Clearing House must be fully functional when the Protocol will enter into force. Therefore it is important that it is developed in close collaboration with already existing information-sharing systems on biotechnology and biosafety like the BINAS[3] system of UNIDO or the OECD's BIOTRACK.[4]

[3]http://www.binas.unido.org/binas/.

[4]http://www.oecd.org/ehs/service.htm.

Capacity building: a fundamental requirement

Almost all industrialised countries have already in place a domestic regulatory framework to deal with the use of LMOs. This is obviously not the case in most developing countries. Therefore the development and strengthening of human resources and institutional capacities in biosafety is a fundamental requirement for the effective implementation and operation of the Protocol. The Protocol builds on existing national, regional and international initiatives. It will therefore be important to explore mechanisms to promote the participation of all relevant institutions and initiatives including the private sector in the biosafety capacity-building activities and to insure co-ordination and co-operation among them. Special emphasis shall also be given to insure availability of adequate financial resources. In this respect a stronger support of the Global Environmental Fund (GEF) will be needed.

Future work and challenges

The role of the ICCP (Intergovernmental Committee for the Cartagena Protocol)

The Cartagena Protocol shall enter into force 90 days after the 50th ratification by a State that is Party to the Convention. The first Meeting of the Parties to the Protocol will take place in conjunction with the next Conference of the Parties to the Convention of Biodiversity that is scheduled after the date of entry into force of the Protocol. This would be at the earliest May 2002, the date of the 6th Conference of the Parties to the Convention.

The ICCP was established by the Conference of the Parties to the Convention on biological diversity to prepare the first meeting of the Parties to the Protocol and to facilitate the ratification process. The first meeting of the ICCP will take place in December 2000 in Montpellier, France. The ICCP will focus on issues that the Protocol stipulates must be considered by the first meeting of the Parties and on issues that are central to the operation of the Protocol like:

- information-Sharing: Modalities of the operation of the Biosafety Clearing House
- capacity-building
- liability: process to elaborate rules for liability and redress
- documentation: detailed requirements regarding documentation accompanying LMO-FFP, in particular unique identification mechanism.

Most of those issues have already been discussed previously in this paper with the exception of liability.

During the negotiation, the question of liability and redress for damage resulting from transboundary movement of LMOs was certainly one of the most contentious issue. Already in Cartagena, the majority of countries were in favour of postponing further discussion on liability after the entry into force. This option was integrated in Article 27 of the Protocol which requires that the first Meeting of the Parties of the Protocol shall adopt a process for the elaboration of international rules in the field of liability within a time period of 4 years. This means that the question of liability will certainly occupy the political scene with regard to the further development of the Cartagena Protocol. This should be seen in the broader context of the discussions regarding the development of national and international environmental liability regime. Hopefully it will allow expanding in the area of biotechnology the recent success achieved in the area of transboundary movements of dangerous waste with the adoption of the Basel Protocol on liability and compensation.

Conclusions

The Protocol of Cartagena is the first global regulatory framework in the field of biotechnology and biosafety. The protocol offers a global and flexible framework to respond to the safety needs associated with the international movement and trade of LMOs, especially transgenic crops, which will most probably significantly increase in the next years. Considering that many developing countries do not have biosafety legislation, the Protocol will promote and strengthen the International co-operation between North and South countries in the field of technical and institutional capacity building related to biosafety. The Advance Informed Agreement procedure establish a transparent mechanism which shall allow importing countries to make a sound environmental risk assessment and to decide upon the import of LMOs .The Protocol establishes the Biosafety Clearing house which will serve as the International Information Sharing system on biosafety and biotechnology. By recognising the right for a clear identification mechanism for all products containing LMOs, the Protocol will support the development of international standards to facilitate the traceability of LMOs from the seed producer to the final consumer through all the agro-food chain. Last but not least the concept of mutually supportiveness which set the relation between the Cartagena Protocol and the other international agreements, shall serve as a reference to clarify the relation between the WTO and other Environmental International Agreements and more generally between trade and environment.

Session 3: Impact of biotechnology on conservation

Methods for Risk Assessment of Transgenic Plants
IV. Biodiversity and Biotechnology
K. Ammann, Y. Jacot and R. Braun (eds)
© 2003 Birkhäuser Verlag Basel/Switzerland

The conservation of genetic diversity: gene flow from agriculture

Alan Gray[1], Roger Daniels[1], Alan Raybould[1], Ian Cooper[2], Lindsay Maskell[1], Denise Pallett[2], Mary-Lou Edwards[2], Milo Thurston[2] and Matthew Alexander[1]

[1]*Centre for Ecology & Hydrology, CEH Dorset, Winfrith Technology Centre, Dorchester, DT2 8ZD, UK*
[2]*CEH Oxford, Mansfield Road, Oxford, OX1 3SR, UK*

Introduction

One of the most frequently raised concerns about the introduction of genetically modified crops is that transgenes may be transferred to wild relatives, resulting either in the disruption of natural patterns of genetic diversity by "pollution" of species gene pools or in the addition of traits which may cause wild plants to become weedy or invasive. This paper examines these ideas from the viewpoint of conservation genetics. We look first at what is known about genetic diversity in natural populations and the forces that shape it, and then consider some of the issues which arise from hybridisations between crops and their wild relatives. The problems of predicting which traits may confer weediness and the importance of taking a case-by-case approach are illustrated by some of our work on virus-resistance in wild *Brassica* species.

Genetic diversity in natural populations

Modern reviews of genetic diversity in plant populations, mainly based on an analysis of genes which are selectively neutral, usually overlook the wealth of studies, spanning 80 years, which have unequivocally demonstrated the relationship between genetic and environmental variation. Beginning with the early genecologists, Turesson in the 1920s [Turesson, 1922], Clausen and his associates and Gregor in the 1930s and 1940s [Clausen et al.,1940; Gregor, 1996], and extending to modern studies of population differentiation in response to environmental stresses such as heavy metals [Bradshaw & Mc Neilly, 1981], salinity [Kik, 1989], herbivores [Mc Kinney and Fowler, 1991], herbicides [Powles & Moltum, 1994] and pathogens [Burden, 1987], numerous empirical studies have shown that genetic diversity in most plant species lies along the axis of environmental, and specifically habitat, variability. The widespread occurrence of such genotype-environment correlations provides "a testimony to the power and ubiquity of natural selection" [Gray, 1996]. The force of selection is easily overlooked by the more recent preoccupation with marker genes, the popularity of a reductionist approach to plant ecology, and the rejection of an adaptionist viewpoint [Gould & Lewontin, 1979].

What marker genes, and notably allozymes, have shown us is that interacting with the species' response to selection, and influencing genetic structure, are two main sets of factors: those stemming from its intrinsic biology (the genetic recombination system including the breeding system, pollination and seed dispersal modes, and life form) and those resulting from extrinsic, dynamic processes such as fluctuations in population size, bottlenecks, invasions and many other historical and contemporary events. These latter include modern agricultural practices, which can influence population size and fragmentation. It has been argued elsewhere [Gray, 1996; Gray, 1996a] that, within broad phylogenetically-related groups of plants, a species' genetic system is an extremely useful predictor of the patterns of population differentiation which it is likely to display.

The species' genetic system will determine the nature of its response to selection. Nevertheless, we may feel confident that the optimum way to sample the genetic diversity within a species is to sample the range of habitat-types in which it is found (e.g. grazed meadows, ungrazed meadows, sand dunes, woodland clearing, etc.), and to do so across as broad a geographical and altitudinal range as possible (capturing differences in response to regional climate and resulting from restricted gene flow between populations under increasing isolation by distance). Similarly, any strategy which aims to conserve the full range of genetic diversity in a species should seek to protect representative natural populations across this spectrum of habitats.

Gene flow from agriculture

How might the transfer of transgenes to wild relatives affect the patterns and amounts of genetic diversity in natural populations? Perhaps the first point to make is that this question can only be asked of a rather small subset of all wild plant species. Although nearly all crops may hybridise with a wild relative in some part of their range [Ellstrand et al., 1999], a tiny fraction of the world's flora has been domesticated, and in modern agriculture most crops are grown outside the geographical range of the wild relatives with which they might hybridise. Thus, for example, in the UK the issue of hybridisation does not arise for crops such as wheat (which in 1996 covered 42% of the area of arable crops in England and Wales), maize, potatoes and many others. The extent to which a particular transgenic crop may be potentially be involved in gene transfer will vary from country to country and region to region. In Europe, an analysis of the potential for gene flow to wild relatives has been made in detail for The Netherlands [de Vries et al., 1992], Switzerland [Jacot & Ammann, 1999] and the UK [Raybould & Gray, 1993]. Such analyses, based on variation in natural floras, should be carried out for all countries where genetically modified crops are likely to be grown.

Secondly, during the long process of domestication of many crops they will have hybridised with wild antecedents and relatives, and *vice versa*, such that crop and wild relative share many genes by descent. For example, the cultivated and wild forms of pearl millet, *Pennisetum glaucum*, have exchanged genes for at least 3,000 years, co-evolving over large parts of the Sahel [Renno et al. 1997]. In such cases of common ancestry, or where crops are part of crop-weed-wild species complexes, assessing the likelihood and amount of contemporary gene flow becomes extremely difficult. Furthermore, in cases such as oilseed rape, where the wild relatives are mainly weedy species of agricultural and disturbed land, and where there has been a long history of the establishment of feral populations, and even naturalisation, the patterns of genetic diversity are likely to have been strongly affected by past agriculture. The widespread exchange of seed will also have influenced current patterns of genetic diversity. For example, Figure 1 shows the import of clover into Britain during the eighteenth and nineteenth centuries, indicating that there has been a very long history of importing genotypes from a wide range of sources including, as long ago as the beginning of the nineteenth century, seed from North America. Other agricultural species used for fodder will have similarly been exposed to imported genotypes, probably over many centuries [Ambrosoli, 1997]. Under such circumstances, it is likely to be difficult to define an indigenous genotype, and, as Daniels and Sheail (1999) [Daniels & Sheail, 1999] argue, difficult to sustain the idea that gene flow between crop and feral or wild relative constitutes genetic "pollution".

Figure 1. The import of clover seed to Britain from different sources between 1700 and 1870. Values are tonnes per year averaged over ten-year periods.

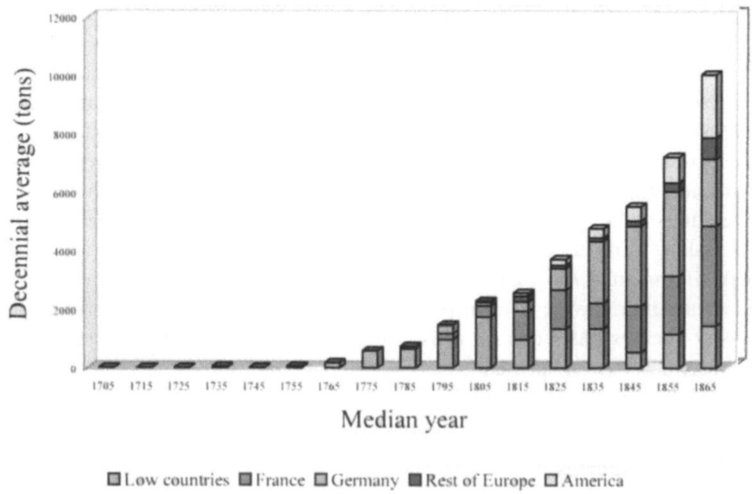

Gene flow from crops can therefore be seen as one of several forces alongside changes in agricultural practice and habitat fragmentation that may shape the genetic diversity present in wild populations of interfertile relatives. In rare cases, such introgression may locally threaten native species, as in the example of gene flow from lucerne (*Medicago sativa*) to tetraploid populations of sickle medic (*M. falcata*) in regions of Switzerland. Here, *M. falcata* has in some regions become endangered as a "pure" species by persistent hybridisation with the crop [Al Mazyad & Ammann, 1999]. In most cases, however, gene flow may be difficult to detect. Evidence of past gene flow may be inferred indirectly, as in ryegrass (*Lolium perenne*) populations in southern Britain, where, in contrast to the patterns of genetic variation in bristle-leaved bent grass (*Agrostis curtisii*), a non-agricultural species with a similar genetic system (outbreeding, perennial), the pattern of genetic variation is consistent with the past occurrence of large-scale gene flow from ryegrass cultivars [Warren et al. 1998]. The "natural" patterns of variation, such as those in *A. curtisii*, where populations from adjacent regions were more similar genetically than those from greater distances, were apparently disrupted in ryegrass by the introduction of non-local genotypes.

Although the ryegrass study employed allozymes to detect variation patterns, the results imply that the genes concerned, or genes to which they were linked, do not negatively affect the fitness of those plants to which they are transferred. Similarly, for transgenes to spread from modern crops into wild and weedy relatives, the key consideration is how such genes affect plant fitness.

Weeds and weediness

A frequently raised concern about the introduction of genetically modified crops is the possibility that transgenes may enable wild relatives to become weedy or persistent. For the transgenes to spread, they would have to be at least selectively neutral; to increase weediness or persistence, they would have to increase fitness. It is not possible to define fitness in absolute terms since it is context-specific - *i.e.* the trait for tolerance to the herbicide glyphosate will increase the fitness of one plant relative to another plant without the trait in situations where glyphosate is applied to both plants. To understand the potential of a particular trait to increase weediness, we must be aware of all those situations in which it may increase relative lifetime fitness.

It is not possible to say of a particular trait that it will always increase (or decrease) fitness, despite the fact that some traits (e.g. seed dormancy, short juvenile phase) are frequently associated with successful weeds. Such traits, sometimes referred to as "Baker traits" (after H G Baker, who first listed them [Baker, 1965]) are only weakly predictive of the weediness of a particular species. In one study of British weeds, the correlation between the number of Baker traits and species weediness was positive but was best described by a curvilinear relationship such that the more Baker traits a species possessed above four or five, the less weedy it appeared to be [see Williamson, 1996]. Clearly, the effects of adding a trait to a wild plant are not easily predicted in a generic way. The importance of detailed specific investigations on a trait-by-trait, species-by-species basis can be illustrated by the example of virus-resistance and its potential impact on wild *Brassica* species.

Virus-resistance in *Brassica* species

One possible effect of the transfer to a wild relative of a "fitness" trait, such as the ability to tolerate a pest or pathogen, is that reduction of the impact of the pest will enable populations of the plant to grow at a faster rate than when controlled by the pest. For this phenomenon, known as "ecological release", to happen, the pest or pathogen must be regulating the population size or growth rate. It is not sufficient to demonstrate that a particular pathogen causes or increases mortality - the effect must be translated into a reduction in population growth. For example, higher mortality at the seedling stage among susceptible genotypes may be offset by greater density-dependent competition between resistant genotypes at a later stage in the lifecycle. Similarly, higher fecundity of resistant populations may not be translated into increased survivorship in the next generation if the main limiting factor is seedling survivorship, or microsite availability. It is therefore important to assess the extent to which particular pests regulate particular species' populations. This is not a trivial undertaking.

We have been investigating the impact of plant viruses on natural populations of wild cabbage (*Brassica oleracea*) and black mustard (*B. nigra*) [e.g. Raybould et al., 1999; Maskell et al., 1999; Raybould et al., 1999a]. It is too early to draw firm conclusions about the possible impact of transferring a virus-resistance gene (from a cultivated *Brassica*) to either of these species, but some very clear trends are emerging from the work (which is centred on populations of both species distributed along the coast of Dorset in south-west England).

First, several viruses are common in populations of both species and many plants are infected with two or more viruses. For each species there is significant variation between populations in the proportion of individuals with a particular virus, and significant between-year variation in the frequency of some viruses. However, there are consistent significant differences between species in virus infection. In *B. oleracea*, large numbers of plants (between 85% and 25%) in most populations were infected with cauliflower mosaic caulimovirus (CaMV), beet western yellows luteovirus (BWYV) and/or turnip mosaic potyvirus (TuMV), whilst turnip yellow mosaic tymovirus (TYMV) was relatively rare apart from one population. In contrast, CaMV (the most common virus in *B. oleracea*) was rare or absent in most populations of *B. nigra*, as was TuMV, apart from a 25% infection in one population in one year, and, to a lesser extent BWYV, which was absent or at low frequency in most populations (TYMV was also relatively rare in *B. nigra*). However, two viruses which were absent from *B. oleracea*, turnip rosette subemovirus (TRosV) and turnip crinkle camovirus (TCV), were commonly found in *B. nigra*, with TRosV infecting more than 40% of plants in three out of five populations in one year. Differences in virus infection occurred even where populations of the two *Brassica* species grew close together (e.g. TuMV present in *B. oleracea* but absent in *B. nigra*, *vice versa* for TRosV).

Secondly, we are beginning to build a picture of the variation in this system in virus pathology and its impact on natural population dynamics. The picture is complex and essentially virus- and species-specific. For example, in a series of glasshouse, common garden and field trials [e.g. Maskell et al., 1999], plants of *B. oleracea*, when compared with controls, showed significantly increased mortality, reduced growth and lower seed output when inoculated with TYMV, higher mortality and lower seed output but similar dry weight when inoculated with TuMV, but no reduction in performance or fecundity when inoculated with CaMV. Interestingly, the severity of impact of these three viruses is ranked in order of their rarity in the populations studied. When plants of *B. nigra* are inoculated with TuMV or CaMV, both rare in natural populations, they almost always die before flowering. The common *B. nigra* viruses TRosV and TCV produce recognisable symptoms and, whilst they do not usually kill plants in the glasshouse, early results indicate that these viruses are a major cause of seedling mortality in the field.

The next stage, that of evaluating the impact on population growth rates of a particular virus on each species in the field has begun and will enable more accurate and specific predictions of the potential effects of transgenes conferring virus-resistance. Already, however, we can be confident that the fitness advantage produced by the acquisition through hybridisation of a transgene will be strongly dependent on the virus and the crop. Thus, a gene providing resistance to TuMV is postulated to have a higher selective advantage in *B. nigra* (in which the virus has killed all the plants so far tested) than in *B. oleracea* (which has some tolerant genotypes in the wild). Ecological release due to this transgene may be more likely in the former.

Indeed, our findings point to the likely importance of factors such as the history of exposure to, and co-evolution with, viruses of wild plant relatives. For example, the most widespread virus in *B. oleracea*, CaMV, has little effect on that species but is a severe pathogen in *B. nigra*, in the natural populations of which it is rare. The impact of TuMV on *B. nigra* also shows that failure to detect a virus in the wild should clearly not be taken as indication that a gene for resistance to that virus will have no impact on feral populations of wild relatives - an argument which has been used in assessing the risks of releasing a transgenic virus-resistant squash in the USA [Kling, 1996].

References

Al Mazyad PR, Ammann K (1999) Biogeographical assay and natural gene flow. In: Ammann K, Jacot Y, Simonsen V, Kjellson G (eds.) *Methods for risk assessment of transgenic plants. III. Ecological risks and prospects of transgenic plants*, 95-98. Birkhäuser Verlag, Basel

Ambrosoli M (1997) *The wild and the sown. Botany and agriculture*. 1350-1850. Cambridge University Press, Cambridge

Baker HG (1965) Characters and modes of origin of weeds. In: Baker HG, Stebbins GL (eds.) *The genetics of colonising species*, 147-172. Academic Press, New York

Bradshaw AD, McNeilley T (eds.) (1981) *Evolution and Pollution*. Arnold, London

Burden JJ (1987) Phenotypic and genetic patterns of resistance to the pathogen *Phakapsora pachyrhizi* in populations of *Glycine canescens*. *Oecologia* 73: 257-267

Clausen J, Keck DD, Hiesey WM (1940) Experimental studies on the nature of species. I. The effect of varied environments on western North American plants. *Carnegie Institute of Washington Publications* No 520. Washington DC, 1-452

Daniels RE, Sheail, J (1999) Genetic pollution: concepts, concerns and transgenic crops. In: Lutman PJW (ed.) *Gene flow and agriculture, relevance for transgenic crops. BCPC Symposium No 72*, 65-72. British Crop Protection Council, Farnham de Vries FT, van der Meijden R, Brandenburg WA (1992) Botanical files. A study of real chances for spontaneous gene flow from cultivated plants to the wild flora of The Netherlands. *Gorteria* supplement, Rijksherbarium Leiden

Ellstrand NC, Prentice HC, Hancock JF (1999) Gene flow and introgression from domesticated plants into their wild relatives. *Annual Review of Ecology & Systematics* 30: 539-563

Gould SJ, Lewontin RC (1979) The spandrels of San Marco and the Panglossian paradigm: a critique of the adaptionist programme. *Proceedings of the Royal Society of London B* 205: 581-598

Gray AJ (1996) Genetic diversity and its conservation in natural populations of plants. *Biodiversity Letters* 3: 71-80

Gray AJ (1996a) Genecology, the genetic system and the conservation genetics of uncommon British grasses. In: Crawford TJ, Spencer J, Stevens D, Tew TE, Usher MB, Warren J (eds.) *The role of genetics in conserving small populations*. Joint Nature Conservation Committee, Peterborough

Gregor JW (1996) Ecotypic differentiation. *New Phytologist* 45: 254-270

Jacot Y, Ammann K (1999) Gene flow between selected Swiss crops and related weeds: risk assessment for the field releases of GMOs in Switzerland. In: Ammann K, Jacot Y, Simonsen V, Kjellson G (eds.) *Methods for risk assessment of Transgenic plants. III. Ecological risks and prospects of transgenic plants*, 99-108. Birkhäuser Verlag, Basel

Kik C (1989) Ecological genetics of salt resistance in the clonal perennial *Agrostis stolonifera*. *New Phytologist* 133: 453-458

Kling J (1996) Could transgenic crops one day breed superweeds? *Science* 274: 180-181

Maskell LC, Raybould AF, Cooper JI, Edwards M-L, Gray AJ (1999) Effects of turnip mosaic virus and turnip yellow mosaic virus on the survival, growth and reproduction of wild cabbage (*Brassica oleracea*). *Annals of Applied Biology* 135: 401-407

McKinney KK, Fowler NL (1991) Genetic adaptations to grazing and mowing in the unpalatable grass *Cenchrus incertus*. *Oecologia* 88: 238-242

Powles SB, Holtum JAM (eds.) (1994) *Herbicide resistance in plants*. Lewis Publishers, London

Raybould AF, Gray AJ (1993) Genetically modified crops and hybridisation with wild relatives: a UK perspective. *Journal of Applied Ecology* 30: 199-219

Raybould AF, Moyes CL, Maskell LC, Mogg RJ, Warman EA, Wardlaw JC, Elmes GW, Edwards M-L, Cooper JI, Clarke RT, Gray AJ (1999) Predicting the ecological impacts of transgenes for insect- and virus-resistance in natural and feral populations of *Brassica* species. In: Amman K, Jacot Y, Simonsen V, Kjellson G (eds.): *Methods for risk assessment of transgenic plants. III. Ecological risks and prospects of transgenic plants*, 3-15. Birkhäuser Verlag, Basel,

Raybould AF, Maskell LC, Edwards M-L, Cooper JI, Gray AJ (1999a) The prevalence and spatial distribution of virus infections in natural populations of *Brassica oleracea*. *New Phytologist* 141: 265-275

Renno JF, Winkel T, Bonnefous G, Bezançon G (1997) Experimental study of gene flow between wild and cultivated *Pennisetum glaucum*. *Canadian Journal of Botany* 75: 925-931

Turesson G (1922) The genotypical response of the plant species to the habitat. *Hereditas* 3: 211-350

Warren JM, Raybould AF, Ball T, Gray AJ, Hayward MD (1998) Genetic structure in the perennial grasses *Lolium perenne* and *Agrostis curtisii*. *Heredity* 81: 556-562

Williamson M (1996) *Biological invasions*. Chapman & Hall, London

Methods for Risk Assessment of Transgenic Plants
IV. Biodiversity and Biotechnology
K. Ammann, Y. Jacot and R. Braun (eds)
© 2003 Birkhäuser Verlag Basel/Switzerland

Problems of plant conservation in agricultural landscapes: can biotechnology help or hinder?

Brian Johnson

Biotechnology Advisory Unit, English Nature, Taunton TA1 5AA, UK

Introduction

Plant gene pools, whether in crops or wild plants, are probably disappearing faster now than they have since the mass extinctions found in the fossil record. Habitat destruction is the main cause, with the global expansion of intensive agriculture a major factor. Besides the disruption and destruction of ancient and irreplaceable ecosystems, this loss of genetic material should be of major concern to any scientist involved in biotechnology and other forms of plant breeding.

Biodiversity is an important component of farmland landscapes because:

- There are many native species that are entirely dependent on farmland at critical stages in their life cycles. This may be because they have co-evolved with agriculture or because habitats they once occupied have been replaced by agriculture.
- Biodiversity within agriculture can act as an early warning system if agricultural methods are potentially unsafe to humans. There are some classic examples of cases (such as organochlorine pesticide use) where bio-accumulation within native species in contact with agriculture has revealed toxicity which was not identified in standard safety tests (Mellanby, 1967; Sheail, 1985)
- Pest control and avoidance, especially in organic agriculture, may partly depend on the presence of biodiversity within and around crops. Populations of predatory insects and birds are important factors limiting damage to crops by economically harmful insects.
- Biodiversity is seen by society as an indicator of sustainability and good stewardship of farmland. The Convention on Global Biodiversity is an expression of worldwide concern about reductions in biodiversity, and the maintenance of biodiversity is a key part of the definition of sustainability agreed in the Kyoto Convention. Within Europe, the Habitats and Species Directive and the Birds Directive place obligations on member states to conserve biodiversity. This includes many farmland-dependent species of plants and animals and farmland habitats, which are given special protection under the Directive.

Within agricultural landscapes in the past 50 years, there has been a relentless progression from extensive systems of relatively low crop yield and high biodiversity towards highly intensive systems of forage and arable crops, where monocultures with low native biodiversity are normal. This change in agriculture marks the diversion of almost all field biomass towards food production, leaving little for farmland-dependent native biodiversity. Such biodiversity, especially plant diversity, has co-evolved with arable agriculture since the dawn of cultivation, probably since the Neolithic era in the Middle East, where cereals and other crops first developed. It is thought that ecosystems based within arable cultivation are older than some other managed semi-natural habitats such as heathlands and grasslands (Potts, 1997).

The wild plants found within extensive grasslands and cultivated food crops form the basis of important farm-dependent ecosystems that support increasingly rare species of invertebrates, mammals and birds. Wilson (1994) estimated that twenty-five of the 200 species of British arable plants are now Nationally Scarce and a further 24 are of conservation concern. Fifteen species of once-common arable plants are now so threatened that special measures are needed to ensure their survival (Stewart et al., 1994). More effective herbicides are largely responsible for these declines (Wilson, 1992; 1994) and similar trends have been observed elsewhere in north western Europe (see examples in Andreasen et al., 1996 and Eggers, 1984).

Insects associated with non-crop biomass have also declined markedly over the past 40 years (Donald, 1998), to the point where many species are now confined to areas of extensive farming, mostly on nature reserves.

The biomass of farmland birds in the UK is estimated by Dolton and Brooke (1999) to have declined by at least 29% from 1968 to 1988 and the British Trust for Ornithology has presented census data (Fuller et al., 1995 and Siriwardena et al., 1998) showing that populations of at least 13 species of farmland birds are now considered to be so low that special remedial action will be necessary if they are to survive. The fact that bird biomass has declined so much on farmland, strongly suggests that plant and insect biomass available to these species has also declined.

Research in the UK (e.g. Ewald and Aebischer, 1999) has revealed the importance of native plants to farmland birds in arable and grassland fields. Long-term study of the grey partridge *Perdix perdix* in England by the Game Conservancy Trust shows that, for this species, reduced seed and insect availability is a major factor in its decline. When measures are taken on arable fields that allow more native plant biomass to grow within the crop, increasing insects in fields, populations of this species recover to viable levels (Potts, 1986). Similar studies by Green and Griffiths (1994) and Evans (1997) involving experimental manipulation of cropping systems have shown that populations of stone curlews *Burhinus oedicnemus* and cirl bunting *Emberiza cirlus*, species dependent on native plants and insects in arable crops, can be enhanced by changing crop rotations and allowing more in-crop biodiversity to survive. A review by Campbell et al. (1996) of the indirect effects of pesticides on farmland birds revealed the importance of plant biomass to bird populations within cropped landscapes.

Besides reductions in plant biomass and the distribution and abundance of arable plants, there has also been a marked shift in the species composition of weed populations within arable fields. The abundance and diversity of dicotyledonous species appears to have declined in favour of grass weeds such as *Alopecurus myosuroides*, *Agrostis stolonifera* and *Lolium perenne* even within field margins (Firbank, 1999 and Bunce et al., 1999). The cause is primarily the greater susceptibility of dicotyledonous plants to herbicides, although there are some exceptions to this such as *Chenopodium album*, *Stellaria media*, *Galium aparine* and *Matricaria discoidea*; species which have developed resistance to some commonly used herbicides (Stewart et al., 1994). Besides the apparent overall reduction in native plant biomass within fields, this shift in the weed spectrum is thought to be a major factor affecting the decline in insects and birds, since broadleaved weed species generally produce higher seed biomass and appear to support insect groups, which are main components of the diet of birds (Potts, 1997).

This scenario of declining farmland wildlife is typical over large parts of Europe (Krebs et al., 1999) and elsewhere where intensive chemical-dependent agriculture is practised (McLaughlin and Mineau, 1995; Freemark and Boutin, 1995). In Europe there is only a small proportion of wilderness in the landscape, so conservation of many native species depends on farmland management. The issue of declining wildlife in agriculture may be of less concern in large countries such as the US, where intensively farmed land occupies a smaller proportion of the total area, and where large tracts of natural or extensively grazed wilderness act as refuges for many native species.

Factors responsible for declines in rare arable plants and the reduction of plant biomass in agricultural fields

Herbicide use

Perhaps the major cause of the decline in specialised arable and grassland native flora has been the increase in herbicide efficacy over the past 50 years. Figure 1 shows the increase in the number of herbicide formulations, from the first few herbicides in the 1950s to the hundreds of active compounds used in modern agriculture.

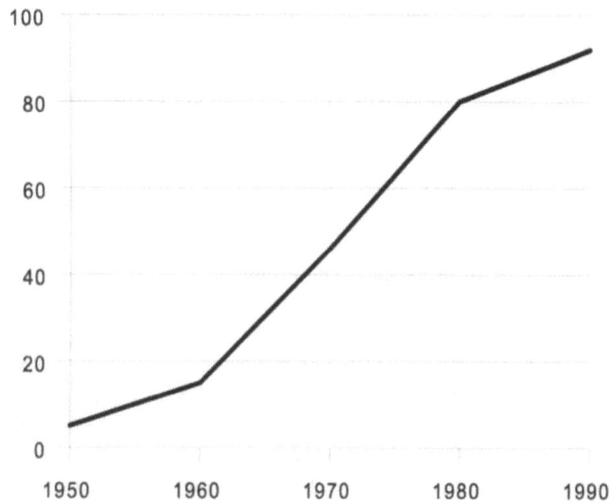

Figure 1. Herbicides approved in the UK (after Hance and Holly, 1990).

The current practice of using pre-emergence, residual and selective herbicides on arable fields, and selective herbicides on grasslands, has led to significant reductions in both the abundance and diversity of native plants associated with arable fields and grasslands. The use of residual herbicides such as the triazines has been especially damaging as these substances suppress germination of weed seeds for much of the growing season. Many farmers now use herbicides prophylactically, aiming at maximising yield and profit.

The use of these herbicide regimes also causes collateral damage to habitats around fields, mainly by overspray and drift. This is thought to be a major factor in the decline of plants and insects found in hedgerows and field headlands in the UK (Boatman, 1989; Marris et al., 1989).

Fertilisers

High levels of fertilisers used on modern crop varieties may also have adverse effects on arable and grassland native plants, either by increasing competition from the faster growing crop or possibly by direct toxicity from phosphates and nitrates (Kleijn and Snoeijing, 1997).

Changes in timing and rotation of crops

One of the main causes of declines in farmland native flora (and therefore fauna) has been the change from spring to autumn sowing in cereals and oilseed rape. There has been a steady increase in the proportion of the year when the land is occupied by crops, leading to elimination of winter fallow and stubbles over large parts of Europe. Winter crops require effective herbicide and fungicide regimes for the crop to succeed, with few weeds being tolerated in the fields. Crop rotations themselves have also changed towards intensive systems where cereals are grown more frequently and fallow fields are no longer part of the system. It is probable that, besides reducing the abundance and diversity of native farmland plants, these modern rotations are leading to significant reductions in the buried seed-bank of arable plants.

The net impact of these changes in agriculture has been to enable farmers to grow crops which occupy the land for longer, give higher yields and be grown in almost weed-free fields, free of many serious pests and diseases. Most farmers would agree that these are highly-desirable aims, but many consumers and politicians are increasingly concerned about the impacts of these methods on biodiversity, landscapes and rural socioeconomic structure.

Key factors in the recovery of farmland-dependent wild flora

The key to improving levels of biodiversity dependent on farmed landscapes is to increase the biomass and biodiversity of the plant trophic level within and around grasslands and food crops, both spatially and temporally. This may be achieved by producing cropping systems which have more wild plants associated with them, and crop rotations that ensure wild plant diversity and biomass within and around crops from season to season. Changing to such systems is a major challenge to agricultural policy-makers, agronomists and farmers, who are beginning to experience deep public concern about the effects of intensive agriculture on nature. The consumer-led movement towards organic production is partly driven by a belief that organic methods are better for the natural environment than conventional chemically-dependent agriculture. There is evidence to suggest that this belief may be justified (see for example Chamberlain et al., 1999 and Kay and Gregory, 1998 and 1999), but that is no reason to accept that organic methods are the only way to mitigate the effects of agriculture on the environment. Given the high demand for food crops fuelled by rapidly rising global populations, it is unrealistic to hope for a return to the biodiversity-rich fields of extensive agriculture, but there may be other ways of increasing plant and animal biomass within cropped landscapes.

Firstly, it is wrong to argue, as some seed companies have done in recent years, that declines are caused by increases in the volumes or application rates of herbicides and pesticides used on crops (including forage crops), and therefore we should be looking for ways to decrease rates and amounts applied. In fact volumes and rates of active substance for some herbicides applied to arable land have been falling since 1984 in the UK (Cooke and Burn, 1995), yet we have seen an increase in the rate of decline in farmland wildlife over that period. Potts (1997) and Campbell et al. (1997) believe that a major cause of this decline is an improvement in the efficacy of agrochemicals, with more herbicides and pesticides becoming available, with better targeting, specificity and timing. If we want to increase native plant biomass and diversity on farmland we should be examining the impact, not input, of herbicides.

There has also been an upsurge in plant breeding, with new crop varieties enabling winter survival, and giving higher yields. It is the enabling effect of new crop varieties that has led to the major shifts in crop rotations described in the previous section. New crop varieties are constantly being developed to extend the range of agronomic options for both forage crops (especially fodder maize and legumes), and food and industrial crops such as potatoes, oil-seed rape and cereals. This is leading to more specialisation and a decline in traditional mixed farms, with their biodiversity-rich mosaics of grassland and arable fields. The development of crop varieties is therefore a key factor to be considered if we want to develop cropping systems that can support greater biodiversity.

It is important to remember that although farmland wildlife has suffered serious declines during agricultural intensification, there are still remnant populations of rare arable plants, although some are on the brink of extinction. Populations of native plants, insects and birds also survive on fields where intensification has not been fully achieved, or on land which has been in organic production or extensive management for a long time. Even in a very intensive crop such as sugar beet in the UK, there is a proportion of fields which can still be classed as weedy (Lainsbury et al., 1999) where farmers achieve much lower levels of weed control than average. There is also still a substantial arable weed seed-bank in many agricultural soils. It is not too late therefore to put in place measures to reverse some of the declines we have seen over the past five decades.

Above all, conservation and enhancement of farmland wildlife will depend on a change in attitude towards biodiversity. This is already occurring among some sectors in the farming community and is embedded within some agricultural policies, but is not yet an integral part of the agrochemicals and plant breeding industries. There have been some welcome moves towards less harmful pesticides, and organic farming has been put on a more commercial basis, but we have yet to see the enhancement of biodiversity as a specific objective in commercial research and development programmes.

Recovery in practice

There are several ways in which the problem of declining farmland biodiversity can be addressed. These can be used synergistically over large farmland landscapes:

Planned biodiversity using field margins and set-aside areas; by which specific areas within fields are deliberately managed to increase biomass and diversity of wild plants. This is most commonly done by having unsprayed margins or strips within a field. Long-term experiments in the UK have demonstrated that declines can be reversed in wild plants and some species of birds by using such methods, and that there are sound agronomic arguments for using planned biodiversity in arable situations, especially where incentive payments via agri-environment schemes are available. A pilot-scale *Arable Stewardship* programme has been developed in the UK, and set-aside provisions have been used to promote less intensive management of arable field margins. Even though there are plans to expand field margin agri-environment measures, the Institute of Terrestrial Ecology (Firbank, 1998) has pointed out that very large areas of suitable habitat would need to be created if declines in farmland birds are to be reversed. This is unlikely to be acceptable in the present agronomic climate. Additionally, some species of farmland birds such as *Vanellus vanellus* require suitable food resources in field centres and crop rotations which increase food availability by the provision of adjacent spring till and pasture (Wilson et al., 1999). Field margin agri-environment measures would be unlikely to benefit such species.

Planned biodiversity using integrated pest and disease control methods; where plant and animal biomasses are increased throughout the field by precise adjustment of crop rotations and agrochemical methods, allowing a greater proportion of wild plants to co-exist within the crop. Examples of these methods can be found in integrated pest control (IPC) programmes where pest and disease control are only carried out when necessary, and at an appropriate level and integrated pest management (IPM) where islands of biodiversity are created to increase predatory insects. During IPC programmes, weed control is only performed at a level to optimise profit and weeds that do not jeopardise this aim are tolerated within the crops. Again long-term studies of these methods at farm-scale have demonstrated that biodiversity can be increased using IPC and IPM, but because IPC techniques are very precise and rely on regular crop monitoring, they are unlikely to be adopted outside farms having access to regular agronomic advice. The driving force behind adoption of these methods is optimisation of profit and yield, rather than increasing biodiversity.

Evolved biodiversity; this is the arable extensification solution in cereal crops favoured by Potts (1997) where plant varieties and cultivation methods are developed which allow greater tolerance to higher levels of biodiversity within the crop. The key to this approach lies not only in lowering pesticide and agrochemical impacts, but also in developing new varieties of crop which can produce their own defences against pests and diseases, reducing the need for pesticides, and crop systems which can also be managed in a way which allows greater wild plant biomass within the crop. Using these methods in the long term could allow the re-establishment of traditional crop-based ecosystems, including achieving the important threshold levels of plant biomass necessary to support higher organisms such as breeding and wintering birds. As Firbank (1999) has pointed out, it is unrealistic to assume that we can return to the very weedy fields of the last century, but we may be able to achieve tolerance of wild plants both within and around the crop.

Unless positive steps to increase within crop biodiversity are taken now, we may witness further declines in arable and grassland wild plants, especially as buried seed banks of these plants become exhausted. Conventional plant breeding may only partly help in the quest for suitable crop varieties for weedier management regimes, because gene pools are limited, even if mutational methods are used.

Can biotechnology help to reduce damage to agricultural biodiversity?

The current generation of transgenic crops is dominated by those modified to be herbicide-tolerant (GMHT crops). Tolerance to broad spectrum contact herbicides (the most usual transformation found in oil seed rape, beet and maize) could be damaging to biodiversity if used to its full potential, but equally could allow more precise control of weeds at the point where they pose a real threat to yield, giving the option of weed tolerance at other times.

Potential of herbicide tolerance to enable damage to biodiversity

If GMHT varieties are used to their full potential, without restrictions implicit in release consents, we can gain some insight into the effects on biodiversity by examining research done at Morley Research Station in the UK. Read and Bush (1998) compared the effectiveness of weed control using conventional and GM glyphosate resistant varieties of sugar beet. Table 1 summarises their results and shows that the GMHT system achieves virtually complete weed control compared with much lower levels using a conventional variety with routine mixtures of herbicides.

Table 1. Mean number of broad-leaved weeds per m^2 remaining after treatment of GM sugar beet with conventional and glufosinate herbicides in 1996

	Dose in l/ha	Weeds per m^2	Yield in tonnes/ha
Conventional	1 x 1.5	25	6.83
Glufosinate	2 x 3	4	7.93
Glufosinate	3 x 2	1	7.53

After: Read and Bush 1998 *Aspects of Applied Biology* 52.

Table 2. Proportion (%) of weeds left after treatment of winter (w) and spring (s) oilseed rape with conventional and glufosinate herbicides

	1995w	1996w	1997w	1995s	1996s	1997s
Conventional	28	13	22	5	53	22
Glufosinate	2	2	5	0	3	7

Data compiled from Read and Ball (1999) *Aspects of Applied Biology* 55.

Later studies by Read and Ball (1999a) produced similar results for GMHT oilseed rape crops. Table 2 has been compiled by estimating means from their experiments with spring and winter oilseed rape crops.

In their research GMHT varieties consistently gave higher degrees of weed control than conventional herbicide regimes in all seasons from 1995 to 1998. There was also greater variation in conventional herbicide effectiveness from relatively low levels of weed control in 1996 to levels as high as those of the GMHT system in 1995. Temporal and spatial variation in the effectiveness of conventional herbicides allows weed growth potentially benefiting biodiversity over time and in patches across farmland landscapes. We can conclude from these experiments that GMHT varieties expressing tolerance to glufosinate will enable much more reliable and complete weed control than conventional varieties. Glyphosate tolerant varieties are likely to show similar results, as glyphosate is at least as effective as glufosinate in controlling most common weeds. Baldwin (1999) shows that farmers in the United States are achieving higher levels of weed control with GMHT varieties of cotton, corn and soybean than they achieved using previous conventional herbicide regimes. Derksen et al. (1999) show that weed control in GMHT canola is better than using conventional herbicides such as imazethapyr and ethylfluralin.

These results indicate that GMHT varieties can enable farmers to remove even more native plant biomass from arable and grassland crops than they are presently achieving. In an already impoverished cropped landscape this ability threatens farmland insect and bird populations even more than conventional herbicide regimes.

Potential of herbicide tolerance to benefit biodiversity

It is possible that the use of HT crop varieties would give farmers confidence that they can deal with weeds when necessary, enabling them to tolerate higher levels of arable weeds than at present. Recent research by Dewar et al. (2000) demonstrates that the timing of broad spectrum herbicide applications is a critical factor affecting both biodiversity and yield within beet fields and shows that it may be possible to enhance biodiversity under some circumstances, without paying an unacceptable yield penalty. For these methods to be adopted would require a change in attitude from arable farmers, perhaps encouraged by agri-environment incentives. This research and on-going studies on GMHT maize in the UK raises the possibility of a return to under sowing of cereals and maize giving agricultural advantages of autumn catch grazing, and enabling a return to winter fallows that benefit biodiversity.

In certain crops, especially those where triazines and other residual herbicides are used, transgenic crops with tolerance to broad-spectrum herbicides may offer considerable environmental benefits. Read and Ball (1999b) show that weed control using glufosinate tolerant maize was as effective as using atrazine-based systems on conventional varieties. In fodder maize, GM varieties with glyphosate and glufosinate tolerance could replace traditional atrazine treatment, lowering risks of acquifer pollution and with the potential to allow higher native plant biodiversity to develop within the crop later in the growing season.

What could be the most likely effects of the current generation of GM crops on botanical biodiversity?

Having discussed the potential for either harm or benefit from transgenic (mostly HT) crops, it is reasonable to speculate on whether the potential benefits to biodiversity will be realised, or whether the risks will materialise, given the current regulatory system and commercial market for these crops.

For the past decade the strategic direction of the use and promotion of HT, whether transgenic or conventionally bred, has been towards greater herbicide impact within most crops, offering a greater degree of weed control, more easily achieved, and at lower cost. The results from the Read *et al.*, experiments show that farmers may have good reason to adopt HT varieties. Experience in the US has shown that GMHT is a very powerful and effective method of achieving very high and reliable degrees of weed control, especially on crops and soils where weed control is difficult.

The British conservation agencies believe that farmers would be most likely to use GMHT varieties in similar situations in the UK, especially where fields are still relatively weedy and where farmland biodiversity persists despite the best endeavours of farmers and agronomists. These relatively species-rich areas would be the primary target for marketing HT crops, risking yet further damage to arable and grassland native flora. This of course assumes that, as in the US, GMHT varieties are released with no mitigatory conditions attached to consents, and that consent is given for farmers to use the herbicides in the most agronomically effective way. This may not be the future for Europe, where regulations are coming into force to encourage regulators of GMHT crops and herbicides to take into account potential effects on biodiversity.

The UK government has started work on a three-year series of field scale experiments to assess the potential effects of using GMHT varieties and their herbicide systems, comparing them to typical standard herbicide regimes for oil seed rape, maize and beet. The results from these studies should enable the regulatory system to estimate the effects on biodiversity of using the GMHT system, and to make better informed decisions on whether the crops should be commercialised and under what conditions.

Even if it is found that there are ways of using herbicide tolerance to reduce damage to biodiversity, Hubbell and Welsh (1999) in a review of the structure of the biotechnology industry, conclude that there is little incentive for biotechnology companies to look for more environmentally-friendly and sustainable farming methods. Relatively small amounts of public funding go into the development of GM crops. At present there is little activity in putting together public/private packages consisting of private research and development coupled with public agri-environment incentives to give sustainable profit margins.

There is clearly a need for greater engagement between public institutions and the biotechnology industry to change strategic targets of research and development towards sustainability. Otherwise socially and ecologically important biodiversity targets are in danger of being ignored by the solely market-driven direction of agricultural biotechnology. There is also an important element of foresight here, because the current generation of GM crops were developed within the agricultural paradigms of over a decade ago. With the recent changes in agricultural objectives towards more sustainable methods, it is not surprising that some GM crops are under close scrutiny by regulators who now require more stringent proof of sustainability.

Transgenic crops of the future?

Since the discovery in 1999 by the John Innes Centre that dwarfing genes in *Arabidopsis* are active in crop plants, there is the possibility of development of fast growing, high yielding varieties, with dwarfing transgenes in cereals and possibly other main crops. This would open options of returning to spring sowing, reducing the need for residual herbicides and fungicides, whilst reducing fertilizer inputs and maintaining profitability. Systems based on such transgenic crops might enable a return to winter fallows and stubbles, with benefits to farmland flora, insects and birds.

In the longer term it may be useful to increase research into the genetic basis and gene transfer of allelopathy, the process by which growth of competing plants can be inhibited by substances secreted by plant roots. One can envisage crops which could selectively suppress weed growth, perhaps selecting only those weeds which actually affect profit and yield, leaving harmless plant biodiversity within the field (Grundy et al., 1999).

Given the fact that fertilisers have adverse effects on field plant biodiversity, the advent of crops that are capable of nitrogen fixation should open possibilities of having permacultures of crops that could tolerate native biodiversity.

Combinations of these traits may eventually offer us options of low or no chemical input crop systems with much higher biodiversity than the current unsustainable systems that are currently damaging farmland biodiversity throughout the world. This goal will only be reached if increases in native plant biodiversity are written into research and development programmes as specific objectives, alongside those of better agronomic performance and better food quality.

Gene flow and introgression into native plants

Gene flow from GM crops to wild plants is one of the most contentious issues in the current debate about biosafety in agricultural biotechnology.

Recent research confirms that genes introduced into some genetically modified crops will move into related native species (e.g. Raybould and Gray, 1993; Chèvre et al., 1997 and DETR, 1999). Gene transfer is almost inevitable from crops that have sexually compatible relatives in adjacent natural ecosystems, but not from crops such as the maize and other cereals grown in Europe, whose closest relatives are on the other side of oceans. But should we worry about this?

After all, genes have been moving for many years from conventionally bred crops to wild relatives; in the UK, for example, hybrids occasionally occur between oilseed rape (*Brassica napus*) and native species like wild turnip (*B. rapa*). The difference of course is that genes inserted into GM crops are often derived from other phyla, giving traits that have never been present in wild plant populations. These changes would not be achievable using mutagenesis or other advanced types of plant breeding.

If some of these traits were accidentally introduced to native species gene pools, they could change the fitness (ability to reproduce) and therefore the population dynamics of hybrids between native plants and crops, eventually backcrossing into the native species and becoming established. So the issue is not so much whether the *rate* of gene flow matters (on which there has been copious and rather inconclusive research), but what *impact* specific transgenes might have on native plant ecology and dependent biodiversity (on which there has been very little research). To add genes from other phyla unwittingly and randomly to native gene pools may result in phenotypic effects that change the way entire genomes relate to their physical and biotic environments (Jones, 1999).

With some justification, it has been argued that many transgenes introduced into crop/native hybrids would in fact *decrease* their fitness in the wild, leading to rapid selection of these genes out of the population. This is particularly true of genes designed to prevent germination of saved seed, like the so-called terminator gene. If this were to escape it would commit genetic suicide and not spread into the natural world as has been suggested by some anti-GM campaigners! Lowered fitness might also result from transgenes that impose a metabolic cost on the plant, or increase palatability of plants to their natural herbivores. Transgenes which lower a plant's resistance to disease and pests would also be unlikely to spread within natural populations, unless the rate of gene flow from crop to native species was very high.

But transfer of certain genes, such as resistance to insects, fungi and viruses could *increase* the fitness of any resulting hybrids, possibly forming aggressive weeds or plants that could invade natural ecosystems. Weeds having stacked genes for tolerance to a range of herbicides could also emerge; these would be difficult to control in agriculture. Farmers may eventually need mixtures of herbicides to control such crop volunteers, causing yet more damage to biodiversity. There is already evidence from North America that this multiple tolerance resistance to herbicides is beginning to emerge.

If insect resistance from GM crops was to become established in native plant gene pools, there could be damage to food chains dependent on insects feeding on previously non-toxic wild plants. Not only might there be a direct effect, for many insects are entirely dependent on single plant species, but acquisition of insect resistance in wild plants could change the population dynamics of their relationship with herbivorous and pollinating insects, increasing the risks of them invading agricultural land and natural ecosystems. These risks could also apply to virus and fungus resistance traits. These traits pose even more serious risks for ecosystems in the developing world, where native plant biodiversity is often very high, increasing risks of gene transfer, and the economic and scientific value of ecosystems is often greater. Control of invasive plants generated by transgenic introgression could be a major problem for subsistence farmers and may have implications for biotopes of global importance. Tropical countries also have more wild crop relatives and contact between farmland and natural ecosystems (Ellstrand and Hoffman, 1990) than temperate farmlands.

The science we urgently need to be able to assess these risks is not yet being done. At present we are unable to predict with confidence what impact escaped genes might have on natural and farmland ecosystems. This lack of information is rather disturbing, given the high commercial pressure and rapid timetable for the introduction of GM crops into our landscapes.

Is there a need for genetic isolation in GM crops?

There is clearly a need to set up effective monitoring systems to detect gene transfer and, equally importantly, we need more research to assess ecological impacts. Research in this area would be in the interests of both industry and the environment, but it would be far better for biotechnology companies to produce the next generation of GM crop plants with in-built mechanisms to prevent gene flow.

Perhaps the simplest and certainly the most obvious way to achieve genetic isolation is to make sure that wherever possible plants used for transformations are unrelated to native species and edible crops within the intended market territory. Biotechnology companies and plant breeders should start thinking now about which plants are chosen as platforms for biomedical and industrial product transformations.

Some degree of genetic isolation could also be achieved either by inserting suicidal gene sequences, or by ensuring that pollen incompatibility mechanisms are built into GM crops. There has been some interest recently in the possibility of chloroplastic transformations (Daniell, 1998), based on the assumption that transgenes inserted into chloroplasts would not be inherited paternally. This has been challenged by several workers who argue that there is evidence for paternal inheritance (e.g. Stewart and Prakash, 1998), and that chloroplastic transformations may be very difficult to achieve in crop species. Mechanisms for genetic isolation may not prove to be completely reliable but they could greatly reduce the probability of gene transfer events taking place.

If plant breeders and biotechnologists had been working on the science of genetic isolation many years ago, they might not be embroiled in the gene flow issue now. Should biotechnology become a standard technique for plant breeding, and it is likely that it will, it should be normal ethical practice to minimise gene flow by genetic isolation of crops from the rest of the living environment.

References

Andreasen C, Stryhn H, Streibig JC (1996) Decline of the flora in Danish arable fields. Journal of Applied Ecology 33: 619-626

Aspinall RJ (1988) The changing geography of the common blue butterfly in Northumberland. Entomol. Rec 100:1-6.

Baldwin FL (1999) The value and exploitation of herbicide-tolerant crops in the US. Proceedings of the 1999 Brighton Conference - Weeds, 653-660

Boatman ND (1989) Selective weed control in field margins. Proceedings of the 1989 Brighton Crop Protection Conference – Weeds, 785-794 The British Crop Protection Council, Farnham UK

Bunce RGH, Smart SM, van de Poll HM, Watkins JW, Scott WA (1999) Measuring Change in British Vegetation Centre for Ecology and Hydrology: Grange-over-Sands, UK

Campbell LH, Avery MI, Donald P, Evans AD, Green RE, Wilson JD (1997) A review of the indirect effects of pesticides on birds. Joint Nature Conservation Committee Report 227. Peterborough UK

Chamberlain DE, Wilson JD, Fuller RJ (1998) A comparison of bird populations on organic and conventional farm systems in southern Britain. Biol. Cons. 0: 1-14

Chèvre A-M, Eber F, Baranger A, Renard M (1997) Gene flow from transgenic crops. Nature, 389: 924

Cooke AS, Burn AJ (1995) The environmental impact of herbicides used in intensive farming systems. Proceedings of the 1995 Brighton Crop Protection Conference – Weeds, 603-612 The British Crop Protection Council, Farnham UK

Crick HQP, Baillie SR, Balmer DE, Bashford RI, Beaven, LP, Dudley C, Glue DE, Gregory RD, Marchant JH, Peach WJ, Wilson AM (1998) Breeding Birds in the wider countryside: their conservation status (1972-1996) BTO Research Report No 198 British Trust for Ornithology, Thetford UK

Daniell H, Datta R, Varma S, Gray S, Lee S-B (1998) Containment of herbicide resistance genes through genetic engineering of the chloroplast genome. Nature Biotechnology, 16 (4): 345-348, Department of Environment, Transport and the Regions, 1999 Investigation of Feral Oilseed Rape Populations: DETR Genetically Modified Organisms Research Report No 12, DETR, London

Derksen DA, Harker KN, Blackshaw RE (1999) Herbicide tolerant crops and weed population dynamics in western Canada. Proceedings of the 1999 Brighton Crop Protection Conference – Weeds, 417-424 The British Crop Protection Council, Farnham UK

Dewar AM, Haylock LA, Bean KM, May MJ (2000). Delayed control of weeds in glyphosate-tolerant sugar beet and the consequences on aphid infestation and yield. Pest Management Science 56(4): 345-350

Donald PF (1998) Changes in the abundance of invertebrates and plants on British farmland. British Wildlife 9 (5): 279-289

Dolton CS, Brooke M de L (1999) Changes in the biomass of birds breeding in Great Britain, 1968-88. Bird Study 46: 274-278

Eggers T (1984) Some remarks on endangered weed species in Germany. 7th International Symposium On Weed Biology, Ecology and Systematics. 395-402

Ellstrand NC, Hoffman CA (1990) Hybridisation as an avenue of escape for engineered genes. Bioscience, 40 (6): 438-442

Ewald JA, Aebischer NJ (1999) Pesticide use, avian food resources and bird densities in Sussex Joint Nature Conservation Committee Report, Peterborough, UK

Evans A (1997) The importance of mixed farming for seed-eating birds in the UK. In: D Pain, Pienkowski MW (eds.): Farming and Birds in Europe. The Common Agricultural Policy and its implications for bird conservation. 331-357. London, Academic Press

Firbank LG (1999) The diversity of arable plants - past, present and some futures Proceedings of the 1999 Brighton Conference on weeds, 251-260

Freemark K, Boutin C (1995) Impacts of agricultural herbicide use on terrestrial wildlife in temperate landscapes: a review with special reference to North America. Agriculture, Ecosystems and the Environment, 52: 67-91

Fuller RJ, Gregory RD, Gibbons DW, Marchant JH, Wilson JD, Baillie SR, Carter N (1995) Population Declines and range contractions among lowland farmland birds in Britain. Conservation Biology 9: 1425-1441

Fuller RJ (2000) Relationship between recent changes in lowland British agriculture and farmland bird populations: an overview. In: Aebischer NJ, Evans AD, Grice PV, and Vickery JA (eds.): Ecology and Conservation of Lowland Farmland Birds. British Ornithologists' Union, Tring. pp 5-16

Fuller RJ, Gregory RD, Gibbons DW, Marchant JH, Wilson JD, Baillie SR, Carter N (1995) Population Declines and range Contractions among Lowland Farmland Birds in Britain Conservation Biology 9 (6):1425-1441

Green RE, Griffiths GH (1994) Use of preferred nesting habitat by stone curlews Burhinus oedicnemus in relation to vegetation structure. Journal of Zoology 233: 457-471

Grundy AC, Bond W, Burston S, Jackson L (1999) Weed suppression by crops. Proceedings of the 1999 Brighton Conference on Crop Protection - Weeds, 3: 957-962

Hance RJ, Holly K (eds.) (1990) Weed Control Handbook: Principles. Eighth edition 582pp. Blackwell, Oxford

Hubbell BJ, Welsh R (1998) Transgenic Crops: Engineering a more sustainable future? Agriculture and Human Values 15: 43-56
Jones JS (1999) Almost like a whale: the origin of species updated. p 203 Doubleday
Kay S, Gregory S (1998) Rare Arable Flora Survey 1998. Northmoor Trust UK
Kay S, Gregory S (1999) Rare Arable Flora Survey 1999. Northmoor Trust UK
Kleijn D, Snoeijing GIJ (1997) Field boundary vegetation and the effects of agrochemical drift: botanical change caused by low levels of herbicide and fertiliser. J. Appl. Ecol. 34: 1413-1425
Krebs JR, Wilson JD, Bradbury RB, Siriwardena GM (1999) The second Silent Spring? Nature 400 611-612
Lainsbury MA, Hilton JG, Burn A (1999) The incidence of weeds in UK sugar beet crops during autumn 1998. Proceedings of the 1999 Brighton Conference - Weeds, 3, 817-822
Marrs RH, Frost AJ, Plant RA (1989) A preliminary assessment of the impact of herbicide drift on plant species of conservation interest. Proceedings of the 1989 Brighton Crop Protection Conference – Weeds. 795-802 The British Crop Protection Council, Farnham UK
McLaughlin A, Mineau P (1995) The impact of agricultural practices on biodiversity. Agriculture, Ecosystems and Environment. 55, 201-212
Mellanby K (1967) Pesticides and Pollution. Collins, New Naturalist Series no. 50
Moreby SJ, Aebischer NJ, Southway SE, Sotherton NW (1994) A comparison of the flora and arthropod fauna of organically and conventionally grown winter wheat in southern England. Ann. Appl. Biol. 124: 13-27
Potts GR (1986) The Partridge: Pesticides, Predation and Conservation. Collins: London
Potts GR (1997) Cereal farming, pesticides and grey partridges In: Pain DJ, Peinkowski MW (eds.): Farming and Birds in Europe, Chapter 6. Academic Press
Raybould AF, Gray AJ (1993) Genetically modified crops and hybridisation with wild relatives: a UK perspective. Journal of Applied Ecology 30, 199-219
Read MA, Ball JG (1999a) Control of weeds in genetically modified crops of winter and spring oilseed rape with glufosinate-ammonium in the UK. Aspects of Applied Biology 55: 27-33
Read MA, Ball JG (1999b) The control of weeds with glufosinate-ammonium in genetically modified crops of forage maize in the UK. Proceedings of the 1999 Brighton Conference on Crop Protection – Weeds. 847-852
Read MA, Bush MN (1998) Control of weeds in genetically modified sugar beet with glufosinate-ammonium in the UK. Aspects in Applied Biology 52 401-406
Sheail J (1985) Pesticides and Nature Conservation: The British Experience 1950-1975, Chapter 7, pp103-122, Clarendon Press, Oxford UK
Siriwardena GM, Baillie SR, Buckland ST, Fewster RM, Marchant JH, Wilson JD (1998) Trends in the abundance of farmland birds: a quantitative comparison of smoothed Common Birds Census indices. J. Appl. Ecol. 35: 24-43
Springett JA, Gray RAJ (1992) Effect of repeated low doses of biocides on the earthworm in lab culture. Soil Biol. Biochem. 24 (12): 1739-1744
Stewart A, Pearman DA, Preston CD (1994). Scarce Plants in Britain. Joint Nature Conservation Committee: Peterborough, UK
Stewart CN, Prakash CS (1998) Chloroplastic-transgenic plants are not a gene flow panacea. Nature Biotechnology 16 (5): 401
Tucker GM, Heath MF (1994) Birds in Europe: their conservation status. BirdLife Conservation Series. 3: 366-367 Cambridge: BirdLife International
Wilson PJ (1992) Britains arable weeds. British Wildlife. 3 (3):149-161
Wilson PJ (1994) Botanical diversity in arable field margins. BCPC Monograph No. 58: Field Margins - Integrating Agriculture and Conservation: 53-58

Methods for Risk Assessment of Transgenic Plants
IV. Biodiversity and Biotechnology
K. Ammann, Y. Jacot and R. Braun (eds)
© 2003 Birkhäuser Verlag Basel/Switzerland

Conservation and molecular methods

Hans C.M. den Nijs

Institute for Biodiversity & Ecosystem Dynamics (IBED), University of Amsterdam, Kruislaan 318, 1098 SM Amsterdam, The Netherlands

Introduction

Concern about the future of biodiversity and its importance for earth and mankind has strongly increased over the years past, and conservation biology has developed to a strong field in biology. In studying and developing conservation strategies and rescue methods, biological conservation now has implemented all modern tools and techniques available, including a wide spectrum of biotechnological techniques. These include in vitro propagation, cloning, and cryopreservation in ex situ conservation programmes (in gene banks, for instance; Pence, 1999; Gonzalez-Beinto et al. 1999; Mandal, 1999). Some people expect in the near future genetic engineering to be of help in producing GM races of rare taxa, threatened with extinction, to help these to overcome certain biotic or abiotic risks. For a recent overview of these aspects of biotechnological applications, see Benson (1999). Of course, all kinds of molecular methods are in use now also for the delivery of data on various hierarchical levels of conservation problems. Recent reviews covering much of the subject are by Soltis and Gitzendanner (1999), Haig (1998), Karp et al. (1997), and Smith and Wayne (1996).

In this contribution, I will strongly restrict myself and only show some aspects of the influence, beneficial and/or detrimental, that application of molecular genetics methods have and may further have in future on the field of conservation biology and on biodiversity itself.

The relation between conservation of biodiversity and application of molecular methods can be considered from several points of view. (1) What are the benefits of the application of molecular methods for conservation per se, and (2) what consequences for conservation may follow from the implementation of the new techniques in agriculture? With respect to the latter point, I will very briefly discuss in the second part of this paper a list of potential positive and negative points, successively, that may emerge from the implementation of genetically engineered products (GM plants) in agricultural practice.

Below, I will first give an overview of various aspects of the study and assessment of biodiversity where molecular methods are used, and from which conservation biology is primarily profiting.

Molecular methods applied

A full array of techniques is available now, giving insight in the genetic basis of biodiversity. Depending on the marker chosen, various levels of genetic variation can be studied, providing different clues, at different hierarchical levels of biodiversity. Below, a breakdown of these hierarchical problem levels will be given, reaching from the taxonomic and supra-specific phylogenetic to the within-individual applications. In Table 1, a survey is given of the molecular

Table 1. Molecular methods applied in Conservation Biology. Relative applicability and current intensity of use indicated by X. See text for the various levels and examples of problems addressed.

	Problem level of application		
Techniques	**Taxonomy/Phylogeny s.l.**	**Hybridization**	**Population/Individual**
Allozymes	XX	XXX	XXX
RFLP (1)	X	XX	XXX
RAPD (2)	XX	?	X
AFLP (3)	X	*	X
Mini/microsatellites (4)	XX	XXX	XXX
mtDNA/cpDNA	XXX	XX	XX
Single copy (5)	XX>	XX>	XX>

(1) Restriction Fragment Length Polyporphism
(2) Random Amplified Polymorphic DNA
(3) Amplified Fragment Length Polymorphism : * if species specific bands are available
(4) Multiple Tandem Repeats/Simple Sequence Repeats
(5) Nuclear and mt/cpDNA as well

markers currently used and their most appropriate application to the problem set as indicated by the number of X per position (based on Mace et al., 1996; Haig, 1998; and Harris, 1999).

In conservation biology, allozyme analysis is still frequently used, due to the fact that this technique is rather cheap and markers are co-dominant, which allows for straight forward population genetic studies (Leberg, 1996). Main fields of application now are the population and individual level, but until recently, application on higher integration levels of taxonomy and phylogeny was also common. However, for this purpose, several DNA marker techniques proved to be superior. Population genetic approaches profit strongly from mini- and microsatellite markers (Multiple Tandem Repeats and Simple Sequence Repeats, respectively). These loci behave co-dominant like allozyme loci.

Important for identification of species, for reconstructing phylogenies, and for distinguishing evolutionary lineages are maternally inherited components, mtDNA and cpDNA. Mitochondrial as well as chloroplast genomes are assumed to form maternal lineages, so are widely analysed for phylogenetic studies. There are very many examples available now where information from these markers proved very helpful in identifying conservation priorities. No doubt that with the strong decrease in costs connected to sequencing, single copy (sequence) data will get increasingly important. Full automation of data generation at low costs will make these basic information available for the discipline of Conservation Biology, too, which is by tradition financially relatively limited in resources.

Problem levels, applications and examples

As introduced in Table 1, molecular methods may be usefully applied at several problem levels. By way of a hierarchical approach, I will subsequently discuss some examples and particular aspects (1) in the area of taxonomy and phylogeny solutions, (2) in the problem area of screening for hybrid events of different origin, both within and between species, and, finally, of (3) applications at the population and individual level.

Taxonomy and Phylogeny

Species delimitations and phylogenetic analyses

It goes without saying that identification of species and knowledge on their boundaries is in many cases crucial to conservation. Irrespective of the fact that we for sure never will know all extant species, or perhaps better, all evolutionarily relevant units, pointing to particular species may be necessary, for instance, for law reasons. The USA Endangered Species Act (ESA) is quite strict in allowing only well recognized species for protection. In this respect, mtDNA analyses have clarified species boundaries and status in an increasing number of cases, for instance, in sea turtles and Hawaiian honeycreepers, listed in Haig (1998). Two taxa of USA sea turtles (*Lepidochelys kempi* and *Lepidochelys olivacea*), which for long time were of uncertain specific status proved to be divergent enough at the molecular level to be assigned to the species level, which brought them under ESA protection (Bowen et al., 1993). A similar situation holds for two subspecies of Hawaiian honeycreepers (*Hemignathus virens* ssp. *chloris* and *Hemignathus virens* ssp. *stejnegeri*), which also proved to be so different in mtDNA analysis that full species status could be given, so that protection was much better supported (Haig, 1998). On the other hand, in groups where little doubt exists as to the status of the species, phylogenetic analysis can reveal which species are basal to the group's evolutionary radiation. In *Saintpaulia*, for instance, Möller and Cronk (1997) performed such an analysis and they were able to identify which taxa of the regional species flock were basic and particularly worth to be conserved.

To what extent and how (molecular) phylogenetic data should contribute to conservation decisions is still a matter of fundamental debate. Particularly the quantitative, i.e., statistical evaluation of phylogenetic data relative to the (possibly) uneven distribution of extinction (threats) over taxa, and the degree of phylogenetic diversity in hot spots are under discussion and study, see for a short review of the debate Bininda-Emonds et al. (2000).

Identification of evolutionarily distinct lineages

The above applies to the species level or above, however, in many species infraspecific patterns of variation occur that also may be relevant for conservation. Within the framework of the Convention on Biological Diversity, the definition of biodiversity explicitly includes the variation within species, which not necessarily needs to be expressed in morphology. In this area, quite some debate is going on since it is inherently difficult to tell how much divergence a unit should show to be of importance for future evolution, thus for conservation. Mainly based on maternally inherited DNA, in mitochondria and chloroplasts, analyses have recently shown that many common species are composed of few up to many haplotype lineages that furthermore often show clear geographic patterns. On the basis of these studies, identification of evolutionarily relevant units and lineages is possible, which frequently reflect the past migration routes, as for instance in Europe during the various glacial and interglacial periods. Some recent reviews of these studies are by Newton et al. (1999) and Soltis and Gitzendanner (1999).

Table 2 gives examples of some common tree species showing regional haplotype differentiation. Black alder (*Alnus glutinosa)* and European beech (*Fagus sylvatica)* show remarkable genetic-geographic variation patterns with 13 and 11 haplotypes, respectively. For both species, the patterns suggest that recolonization after the last glacial period predominantly took place from a refuge area in the Carpathian mountains in eastern Central Europe, while in some other Southern regions (Italy, for instance) divergent lineages are still extant that obviously were not able to increase their distribution area. Similar links between patterns of variation and recent recolonization are also found in Japanese beech. Examples from North America are Douglas fir (*Pseudotsuga menziesii*) (Aagaard et al., 1995), and *Liriodendron tulipifera* (Sewell et al., 1996). This sort of information is of large interest for designing (networks of) forest reserves and for preserving genetic resources for reforestation, nursing and eventual breeding. It is becoming more and more evident that very many common herb species also show regional differentiation for haplotype DNA, reflecting their evolutionary history (see Soltis and Gitzendanner, 1999). A phenomenon of special interest for conservation is the occurrence of rare haplotypes within common species. Here, the question arises whether such lineages should deserve separate conservation efforts. This discussion also points to the fact that knowing the various lineages to exist gives no clues yet as to decide which lineage may be of special evolutionary interest over other lineages. Should one protect the rare, the recent, or the genetically remote types (see also Soltis and Gitzendanner 1999)?

Table 2. Tree species showing cpDNA/mtDNA variation reflecting migrational history and evolutionary patterns

Taxon	Region	Haplotypes	Reference
Alnus glutinosa (Black alder)	Europe	13	King and Ferris 1998
Fagus sylvatica (European beech)	Europe	11	Demesure et al. 1998
Fagus crenata (Japanese beech)	Japan	8	Tomaru et al. 1997
Liriodendron tulipifera	USA	2	Sewell et al. 1996
Pinus sylvestris (Scots pine)	Europe	3	Sinclair et al. 1999
Quercus 8 spp. (Oak species)	Europe	23	Dumolin-Lapegne et al. 1997

Regional differentiation patterns have also extensively been studied in animal species. Avise (1992) gives many examples of mtDNA within-species-sequence divergence in a series on SE of USA coastal species belonging to diverse higher taxa. He found clear parallel patterns in, among other species, Black sea bass, Seaside sparrow, Horse shoe crab, and American oyster. All of these showed a marked split between populations from the Atlantic part of the coast as opposed to the Mexican gulf provenances. Similar parallel patterns were found to occur in mainland fresh water fish species of the area. Remarkably, in other taxa such diversification patterns do not occur. Knowledge on these patterns of common descent of population groupings on the one hand, and of regional divergence splits, on the other, is of great value for distinguishing conservation needs within and between species. An example from local scale is given by Parker et al. (1999), who demonstrated the between-population evolutionary divergence in *Poecilopsis occidentalis*, using data on allozymes, minisatellites, mtDNA, and the MHC locus.

Identification of the patterns as discussed above, including the reconstruction of the geographic relations, is commonly known as phylogeography, which basically aims at integrating the regional geographic distribution and the evolutionary history of the pertaining population (complexes). In principle, this approach sheds much light on the within-species differentiation into evolutionary lineages, and on the dynamics of current evolution. It is commonly assumed that such data reveal the evolutionary perspectives with respect to the taxon's future adaptations and possible responses to changes in ecology and/or environment (Avise, 1992; Legge et al., 1995; Moritz, 1994; Francisco-Ortega et al., 1996; Taberlet, 1996; Haig, 1998; Newton et al. 1999; but see also Goldstein et al., 2000).

Since taxonomy based on within-species variation is by definition difficult, special concepts have been developed to accommodate the use of the units distinguished. On the basis of formal phylogenetic and population genetic elaborations of divergence data, two units are proposed: Evolutionarily Significant Unit (ESU) and Management Unit (MU), originally by Ryder (1986), revised by, among other authors, Moritz (1994 and 1995). Box 1 gives the concise description of both concepts, and Figure 1 presents an example based on four populations and a hypothetical allele distribution.

From Figure 1 it will be evident that very thorough knowledge of the pertaining taxa is needed. In practice, this will only rarely be the case for small organisms and the large majority

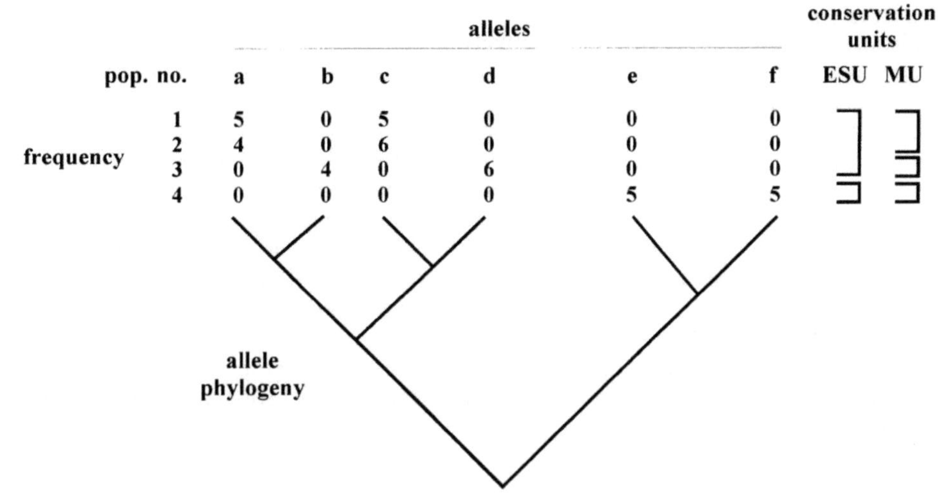

Evolutionary Significant Unit and Management Unit as defined by Moritz (1995)

Figure 1. Evolutionarily Significant Unit and Management Unit as defined by Moritz (1995). Populations 1, 2 and 3 belong to the monophyletic clade containing alleles a, b, c and d, as opposed to the reciprocally monophyletic clade represented by population 4 only. At the local level, the occurrence of alleles b and c in population 3 designates this population to be so different that valuation at MU status follows. Figure redrawn from Moritz (1995).

Evolutionarily Significant Unit – ESU & Management Unit – MU
(sensu Moritz, 1994 and 1995)

ESU:
- reciprocally monophyletic for mtDNA/cpDNA alleles, <u>and</u>
- significant divergence of allele frequencies at nuclear loci

Thus: pattern of evolution, not particularly the amount of divergence

MU:
- significant divergence of allele frequencies at nuclear or mt/cp-loci,
- regardless of the phylogenetic distinctness of the alleles

Thus: population with local adaptations within evolutionary lineages

Box 1.

of plants. This approach indeed is more relevant for larger animal species, eventually involved in captive breeding. However, it might well be useful to assess the relative evolutionary value of, for instance, certain hot spots or assumed diversity centres.

The concept of the ESU and MU gave rise to much debate. To review the discussion would go far beyond the scope of this contribution. A recent and thorough piece is by Goldstein et al. (2000), who on theoretical grounds are very critical to the current practice of the application, but nevertheless conclude that the approach is worthy and should be well elaborated. Independently, Bininda-Emonds et al. (2000) report on a workshop where recently the relevance of implementing ecological data into the concept was stressed upon.

Priority setting among taxa

The applications shown above deliver relatively objective tools for setting priorities for conservation among taxa. Of course, there is a full set of criteria, to name the most important: (1) rate of threat; (2) economic value; (3) ecological role, and (4) biological/evolutionary distinctness. It is at the latter level that recognition of identity by molecular tools may help essentially. Uncovering phylogenetic patterns will also shed light on the areas where recent or current speciation takes place, and therefore gives good indications for the evolutionary position of a given taxon and for the relative value of its related taxa. This includes the analysis of hot spots on the level of species and also of groups of taxa or guilds in other situations. An important assumption in this field is that the patterns revealed indeed reflect the evolutionary potential of the lineages. Given the situation that we have fairly limited knowledge about the concrete relation between the sort of markers in use and the very adaptive value of most of the traits present in the taxa, this assumption is in most cases based on mere guesses. The very strongly developing research field of functional and comparative genomics will likely help underpinning these concepts.

Forensic matters

Among the molecular methods generally most widely known are those providing finger print patterns for identification of individuals or restricted groups. Forensic use for truth finding in all sort of crime cases has been advertised broadly and has been proven very effective, indeed. This also holds for tracking illegal trade in species or products of species that are protected under national law or international conventions, the Convention on International Trade of Endangered Species (CITES) being the most commonly implemented. Because this Convention has many regulations on trade activities on (mostly) the level of species, the availability of species-specific markers is of particular importance. Several technical approaches at DNA level have produced such specific, or even individual, markers, and application in tracking by governments or NGO's of pouching and illegal trade is becoming common practice.

A very illustrative example concerns the analysis of whale and dolphin meat offered for sale on SE Asian (super-) markets. Although commercial trade of the species is either forbidden or very strongly restricted, there appeared to be meat of 18 species in the shops, including a so far unknown Whale subspecies (Ashley, 1999; Baker et al., 1996).

Hybridizations

In conservation biology, hybridization phenomenon may play a role in various ways. Firstly, it may cause problems at the species level when a given species, just by nature, i.e., by historic development, proves to be a hybrid product between other taxa. Secondly, hybridization may help overcome genetic bottlenecks and in re-enforcement programs, particularly with captive breeding. And thirdly, hybridizations may take place between species in more or less natural conditions, with detrimental effects on one or both species involved.

Identifying historical processes

Based on molecular analyses, the status of some species had to be changed. Recently, it has become evident, for instance, that *Eucalyptus graniticola* proved to be a hybrid between two other *Eucalyptus* species that both are rather common. *E. graniticola* was hypothesized to be a very rare and critically endangered relict taxon. Now that its hybrid status has been established, the species is being repleted by hybrid individuals from cultivation (Rosetto et al., 1997). Quite a different and more complicated example, the Red wolf, has been reviewed by Brownlow (1996). This species, seriously threatened in its home range in the SE of the USA also proved to be of hybrid origin, descending from hybridizations between Coyote and Grey wolf. Since the

ESA (USA) does not formally recognise hybrids and, according to the law, the species Red wolf as such does not exist, this would have large implications because the taxon would loose its conservation status, almost certainly leading to extinction. On the other hand, given the current threat to the Red wolf populations, it may be a way out of the genetic problems to re-enforce the weak population with genetic input with one of the parental species. Brownlow (1996) concludes his study by claiming that molecular studies will likely demonstrate more taxa to be of hybrid origin and status. He advocates a discussion to assess the biological value of the phenomenon, and subsequently to implement the findings in the law. Given the situation that hybridization is involved in many plant speciation events, the outcome of this debate may be of large importance for plant conservation too.

Captive breeding

The concept of captive breeding is very well known and applied already in many large animal species. Molecular genetics have enforced these programs strongly. Several marker systems may give insight into the genetic structure of both the population that is in need of re-enforcement and the possible sources. A well elaborated example is the case of the Florida panther (O'Brien, 1994). Here, the program profits from data giving insight into the genetic differentiation that provides possibility to distinguish between more and less relevant units and populations, respectively. A program of genetic re-enforcing the genetically badly depauperate Florida panther is developed, targeting on hybridization with a Texas Puma taxon (O'Brien, 1994). This program is based on comparisons of allozymes, mtDNA and DNA fingerprinting. However, the project is controversial because subspecies boundaries may be at risk and ESA regulations could forbid such deliberate hybridization action.

Another example that may be seen as a warning against undeliberate actions is the case of the Dusky Seaside Sparrow in the SE of the USA. Captive breeding with the last extant individuals by hybridizing these with a morphologically very similar and supposedly phylogenetically close subspecies was unsuccessful, and the rare subspecies went extinct. *A posteriori* DNA analysis revealed that a geographically more remote subspecies was (phylo-) genetically closer, and might have been a better partner taxon for the rescue breeding (Avise and Nelson, 1989).

Analysing current hybridization and introgression processes

With the deployment of the molecular analysis tools, many examples have become apparent where hybridization itself forms a threat to species. In many cases, deliberate or unintentional introductions of alien species into the environment are due to this. Molecular methods may be extremely useful to help tracking alien species or their genetic effects in the form of introgression of certain genome elements. This field reaches from the effects of trials to re-enforce fish stocks, as for instance with Salmonids in USA, to the tracking of introgression from GM races of agricultural crops into wild relatives.

Allendorf and Waples (1995) and Ryman et al. (1995) describe extensive hybridization in Salmonid species, between native and introduced taxa (species, subspecies, races). Introductions were done to restock the regional populations for fishery purposes. Species boundaries are getting fuzzy now, and species are loosing identity. The current hybrid patterns that result from these ill-designed introductions should be sorted out to maintain or to restore viable populations for future.

There are plenty botanical examples, where molecular analyses have discovered hybridizations between common and rare species. Knowledge about such ongoing introgressions helps to restore the genetic identity of the rare species by careful management of the introgressed populations. Table 3 shows some examples of rare *vs.* common species pairs, where introgression is leading to genetic erosion of often small populations of the rare species. Recent reviews list many more examples of this phenomenon, which in fact now is a special concern for rare endemic species (Ellstrand, 1992; Levin et al., 1996; Soltis and Gitzendanner,

1999). Another aspect of molecular methods revealing introgression into wild flora is the analysis of crop-wild relative introgression analysis. Recently, Ellstrand et al. (1999) published a review, showing that the majority of the world's largest crops indeed do have genetic contacts with wild relatives somewhere in the cultivation range, leading to genetic introgression. We will come back to this later. Most frequently, the reported hybridization is due to habitat changes, leading to *de novo* contact zones between the taxa that used to be ecologically separated. Admittedly, hybridization of habitats and secondary contact between species may also lead to new speciation events (Bleeker et al. 1999; Burke et al.1998), but this obviously will only be in a minority of cases.

Table 3. Examples of introgression of common and introduced species into rare relatives. Compiled from Levin et al. (1996), Soltis and Gitzendanner (1999), and Rufener al Mazyad and Ammann (1999).

Common taxon	Rare relative	Area
Arbutus unedo	*A. canariense*	Canary Islands
Cercocarpus betuloides	*C. traskiae*	California
Erigeron lonchophyllum	*E. crispum*	Israel
Gossypium barbadense	*G. tomentosum*	Hawaii
Linaria repens	*L. vulgaris*	UK
Pinguicula grandiflora	*P. vulgaris*	UK
Senecio vulgaris	*S. teneriffae*	Canary Islands
Medicago sativa	*M. falcata*	Europe

The data from the above hybridization studies are of great value for designing reserves, and also for monitoring and management of species integrity, as well as for restoration programs of species and population viability. In the next section, we will elaborate such aspects on the level of population.

Population level problems

Variability within and among populations

Following Harris (1999), assessment of intraspecific genetic variation is important for three prime reasons: (1) the rate of evolutionary change (potential) is directly related to the genetic variation present (Hamrick and Godt, 1996); (2) the level of heterozygosity is positively related to fitness (Allendorf and Leary, 1986; Oostermeijer, 1996 and 2000; Oostermeijer et al., 1995); (3) the gene pool present in earth's biota contains all the genetic information for the biological (and ecological) processes.

As is evident from Table 1, many of the available molecular tools can usefully be applied to this level of conservation problems. However, in conservation genetics in plants, allozyme techniques are still frequently method of choice. Co-dominant traits are far more practical for population genetics, and the costs of allozyme analyses are yet much lower than co-dominant DNA marker techniques (micro-satellites, for instance). On the other hand, allozymes have an obvious disadvantage in that the enzymes rapidly loose activity. Transfer of the material from the often remote and isolated field over larger distances to the laboratory, mostly connected with longer transfer times, is frequently a physical bottleneck. For these reasons, it is to be expected that co-dominant marker systems like microsatellites, and perhaps also sequence data as such, will gain importance strongly in conservation applications.

Analyses of genetic variability mainly aim at assessing the amount and the distribution of the alleles of selected loci within species and within and among populations. In this respect, three important processes can be studied: genetic drift, inbreeding, and gene flow. Important measures are P (proportion of loci polymorphic), A (number of effective alleles per locus), H_o and H_e (proportion of heterozygosity, observed and expected, respectively), and derived values as Fst, a measure for between-population differentiation, signalling the rate of population isolation, for instance. There are many well documented studies published, where the relation between genetic variation and status of the (complex of) populations of a given taxon has been established. The general pattern emerging is that fragmentation and habitat destruction, leading to increased isolation and to a decline of the size of the populations, results in loss of genetic variation, due to a combination of populational genetic processes, reproductive biological phenomena, and demographic events. To name only a few examples, O'Brien (1994) reviewed on mammal species, Ellstrand and Elam (1993) discussed the possible consequences of population genetic processes manifest in small populations particularly for higher plants, and Young et al. (1996) compiled specific examples from field studies. From these and other studies, it emerges that the level of genetic variation is strongly dependent on the type of species concerned, reproductive strategy and breeding system being important factors (Hamrick and Godt, 1989, 1996; Hamrick et al., 1991). In plants, perennial herb species that (1) have mixed-mating reproductive systems (which are normally outcrossing, but are at the same time self-compatible), (2) flower more than once, and (3) that have recently undergone decline in population numbers and sizes, suffer most from the above processes. Particular examples from The Netherlands are given in Table 4 (after Oostermeijer, 2000).

It is concluded that indeed level of polymorphism and effective number of alleles are reduced with population size. The percentage heterozygosity observed is less directly affected. One should be cautious interpreting these results, because most studies are based on extant populations in the field, not on their offspring. Oostermeijer et al. (1995) demonstrated that in the seed population reduced heterozygosity occurs, which the authors led to the conclusion that future generations may show increased homozygosity, and consequently may suffer from inbreeding depression. Basic to this is the expectation that in normally outcrossing species the rate of heterozygozity is an important fitness factor. Indeed, experiments with few rare plant species have established such relations (Oostermeijer et al. 1995 and 1998). In the individuals of *Gentiana pneumonanthe,* the size of the seedlings as well as the number of flowers produced by the progeny are statistically significantly correlated with the heterozygosity rate (Figure 2, from Oostermeijer, 2000). As a consequence, assessment of the inbreeding rate and the analysis of the fitness consequences may be important. This also holds for captive bred reintroduction plant material. Pavlik (1993) carefully prepared reintroduction populations of *Amsinckia grandiflora*, and found an increased level of inbreeding in the garden grown seed families.

Another aspect of assessing the distribution and partitioning of genetic variation among populations is the effect of increasing isolation between the remaining populations of (increasingly) rare species. Fragmentation of originally more or less continuous habitats results in disruption of the metapopulation structure, and therefore negatively influences the between-population gene flow patterns. Molecular markers are extremely helpful in identifying and assessing such effects. The commonly used population genetic measure is Fst, a measure for the amount of variation allocation in the separate populations. It gives an indication of the rate of isolation of the populations, and consequently for the degree of divergence between these. Indeed, rare plants show increased levels of Fst (see Table 4).

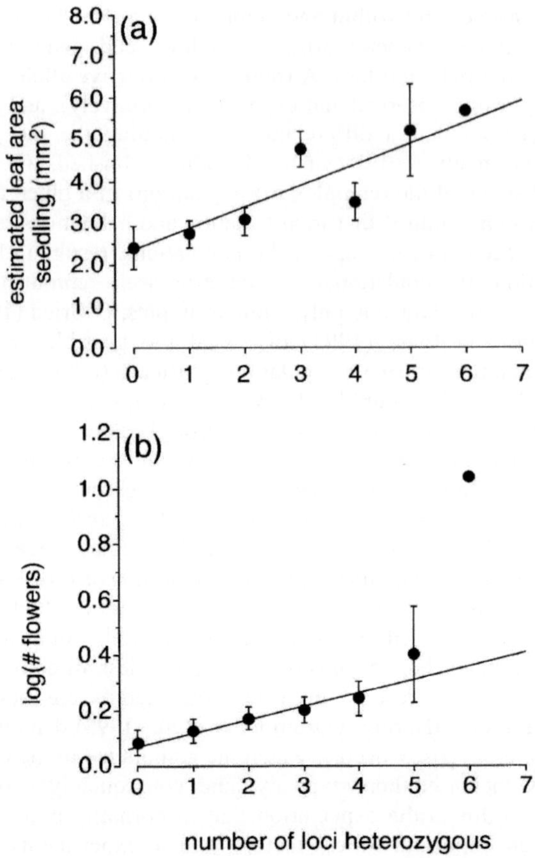

Figure 2. Relationships between individual allozyme heterozygosity for seven variable loci and seedling size (a) and number of flowers produced per plant (b) in Marsh gentian (*Gentiana pneumonanthe*). From Oostermeijer (2000).

Table 4. Relation between population size and parameters for genetic variation. Pearson's product-moment correlation coefficients between ln (population size) and allozyme variation. P = proportion of loci polymorphic, A = mean number of effective alleles per locus, Ho = observed heterozygosity, Fst = the proportion of variation distributed among populations. Significance levels: *: $P \leq 0.05$, **: $P \leq 0.025$. ***: $P \leq 0.001$, (*): 0.05, $P \leq 0.10$, ns: not significant. From Oostermeijer (2000), Oostermeijer et al. (1998); de Boer et al. 1995.

Species	P	A	Ho	Fst
Phyteuma spicatum subsp. *nigrum*	0.161 ns	0.606 *	0.556 *	0.145
Gentiana pneumonanthe	0.491 **	0.332 (*)	0.345 (*)	0.192
Orchis morio	0.004 ns	0.01 ns	0.039 ns	0.162
O. militaris	0.311 **	0.435 **	0.126 ns	0.278
O. purpurea	0.347 **	0.603 ***	0.295 *	0.235
Average for outcrossing perennials				0.094

Population structures
Molecular markers can also be of great value for assessing the intra-population genetic structure of a species. This application is closely related to the above mentioned, and can be seen as an extension, giving a more detailed insight into the genetic status of populations. These insights are very helpful for 1) identification of populations to be re-enforced, 2) identification of possible source populations for re-introductions and re-enforcements, and 3) identification of populations in need of further demographic or environmental consideration (Haig, 1998). Genetic markers are also broadly known for their possibilities for tracking gene flow (as with hybridization, see above).

Genetic population structure may reveal the breeding system of a particular species, and will thus provide clues for its management and its implementation in, for instance, restoration projects. As an example, *Trisetum spicatum* may be mentioned. This tetraploid arctic-alpine perennial grass is used in re-vegetation trials of above timberline ski runs in the Alps (Davos, Switzerland; Urbanska, 1994 and 1995). Multilocus allozyme profiles demonstrated that the natural populations only contain very few genotypes (Table 5A). Further analyses brought to light that the species is predominantly selfing by cleistogamous flowering (Table 5B). It also was concluded that the tetraploid genome of the species showed a high degree of "fixed heterozygosity", due to the presence of duplicated functional loci (den Nijs, unpublished data).

The assessment of these characteristics led to the conclusion that *Trisetum spicatum* is particularly suited for reintroduction and re-vegetation projects, since one needs only few genotypes to start with, and clonal propagation to produce sufficient material does not reduce feasibility of a project.

Knowledge like this should be available prior to the development of reintroduction plans, because it is of large influence on the design and the sampling, and seed and/or plant production protocols (compare Williams and Davis, 1996), who *a posteriori* studied the genetic structure of *Zostera marina* reintroduction populations, and found the level of variation unintentionally and likely detrimentally low).

Population and local race markers
Various molecular techniques provide markers that are population or race specific (mtDNA, RAPD markers, allozyme profiles). As with tracking on the species level, these markers can be very usefully applied to unravel breeding structures in and among populations, the latter clarifying the size and structure of the metapopulation. Analyses of this type were already of great help in developing strategies for conservation of the breeding habitat of aquatic species as the Humpback whale and salmons, as well as for several shore bird species (Haig, 1998).

Table 5A. Multilocus phenotypes and their frequencies in natural populations of *Trisetum spicatum* in Switzerland (den Nijs, unpublished data)

A. Natural populations	**Sample size**	**Number multilocus phenotypes**	**Phenotypes (and frequencies)**									
			1	**2**	**3**	**4**	**5**	**6**	**7**	**8**	**9**	**10**
Chüealptal, GR; site 3	57	3	20	34						3		
Chörbshorn, GR	91	2					63	28				
Grialetsch, GR	27	2				26			1			
Gmeinboden, GR	20	5	1		14				3	1	1	
Scalettapass, GR	24	2		33						22		1
Moiry, VA; site 2	38	2								5		

Table 5B. Field collected progenies of single *Trisetum spicatum* mother plants are completely uniform for multilocus allozyme profile (den Nijs, unpublished data)

Mother/Progenies	Sample size	Number multilocus phenotypes	Phenotypes (and frequencies)									
			1	2	3	4	5	6	7	8	9	10
Chörbshorn #2, mother 26-6	1	1					1					
progeny	15	1					15					
Chüelptal #1, mother 37-11	1	1	1									
progeny	5	1	5									
Grialetsch mother 2-29	1	1							1			
progeny	37	1							37			

Table 6. Within- and between-population gene flow rates in six experimental *Lolium perenne* (CROPPER) populations at two sites [Huizen and Nieuw Vennep], assessed by allozyme marker dispersal. Population size is 20 individuals, except for HU and NV 4, which comprise 60 individuals. Distance between plots as in Figure 3, note that distance between 4 and 5 is 5 meters. From den Nijs and Degenaars (unpublished data).

Plot number		Population size	Total progeny analysed	Donor (HU and NV, resp.)					
				1	2	3	4	5	6
Receptor	HU 1	20	400	97,7	0.5	0.3	0.7	0.5	0.3
	HU 2	20	400	0.0	99.5	0.0	0.0	0.0	0.5
	HU 3	20	400	0.0	0.0	100.0	0.0	0.0	0.0
	HU 4	60	472	0.0	0.4	0.0	99.4	0.2	0.0
	HU 5	20	412	6.8	6.8	0.7	**43.5**	**48.1**	0.2
	HU 6	20	408	0.0	0.2	0.7	1.0	0.5	97.6
	NV 1	20	457	95.2	1.3	0.2	0.4	1.8	1.1
	NV 2	20	409	0.5	98.3	0.0	0.2	0.5	0.5
	NV 3	20	400	0.0	3.0	94.7	1.0	0.8	0.5
	NV 4	60	476	0.2	1.1	1.1	92.4	4.6	0.6
	NV 5	20	426	2.6	4.9	3.7	**35.7**	**50.5**	2. 7
	NV 6	20	425	0.0	0.9	0.7	0.5	0.2	97.2

Population and race specific molecular markers are also very useful for experimental gene flow studies. They are of great value for assessment procedures in biosafety analyses for introduction of GMO's (Klinger and Ellstrand, 1999). With respect to this, modelling gene flow is an extremely important field. Figure 3 and Table 6 give an example of tracking of gene flow between experimental populations of a conventional race of *Lolium perenne* (Cropper), using simple allozyme markers. At two arable sites (HU and NV), avoid of wild *Lolium perenne*, six small populations 20 and 60 individuals, respectively, were planted at set distances (see Figure 3). All individuals per population had an unique allozyme marker, so that influx in any of the other populations could be traced. Regardless of the small size of the populations, gene flow occurred over the small distances set out in the experiment, albeit to a limited level of c. 7% (see Table 6). Obviously deviating from this pattern is the effect of population #4 which contained 60 plants. This difference in size is of particularly large influence on the neighbouring population at a distance of 5 meters. Along the diagonal it is shown that all populations fully set

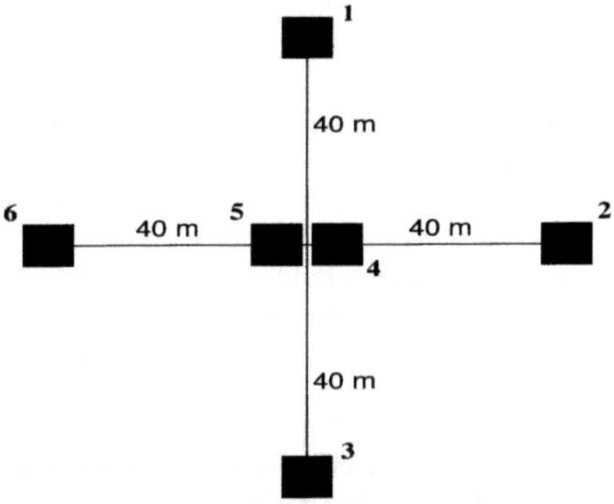

Figure 3. Design of a gene flow assessment experiment between 6 experimental populations of Rye grass (*Lolium perenne*), race CROPPER. The experiment was performed at two sites, Southwest [=Nieuw Vennep] and East [=Huizen] of Amsterdam, respectively (see data in Table 6). Each group of individuals (indicated as populations in the text) has a characteristic (homozygous) allozyme marker. From den Nijs and Degenaars (unpublished data).

seed after within-population pollen transfers, however, the short distance between populations #5 and #4 reduces the within-population pollination rate in the small population (#5) from c. 99% to c. 50% in both experimental sites. In other words, the "large" population #4 born pollen pressure on the neighbouring small one (#5) led to c. 40% effective gene flow into the latter.

This example shows the easiness and value of application of molecular analyses to gene flow studies, but interpreted to the very aim of the pertaining experiment, it also stresses that the relative size of populations is extremely important for the amount of gene flow, proving once more that small populations form clearly sinks for pollen from nearby large populations (den Nijs and Degenaars, unpublished data).

Conservation and the implementation of molecular methods in agriculture

In this chapter, I will briefly review a series of possible effects on biodiversity conservation of the implementation of the new molecular genetics techniques in agriculture. I will mainly discuss two sets of aspects, potentially beneficial and detrimental, respectively. To give an extensive review of the pros and cons would be beyond the scope of this paper, I extensively refer to the publication by Warwick and Small (1999), and to references given by these authors. Of course, there is a plethora of such (more or less likely) consequences, but I will deal only with the conservational aspects of the implementation, not with any other, like effects for agriculture itself, food quantity, quality, or safety.

Potential profits for biodiversity

Most of the reasoning below is self-explaining, although some points are open for criticism. Some of these opposing opinions will be worked out in the next chapter.

Below is a list of likely beneficial implementation aspects.

• *Less stress on natural areas*. It is advocated that if production per unit area will strongly increase due to biotechnology, the need for taking more area in agricultural production will decrease. So, automatically, the pressure on still little used land will decrease, and therefore the biodiversity in such areas will be maintained (Mannion, 1995). Assuming that this hardly will apply to the western high input agriculture, emphasis in this respect is on the developing world's situation.

• *Less use of agrichemicals*. This element is widely advertised and perspectives are positive. Particularly the decrease of pesticides volume may in general be positive for species living sympatric and parapatric with crops. Whether the transgenic traits per se could have positive (or negative) consequences for crop-related weeds and other wild relatives is in some cases still a matter of debate. The same ambivalent situation holds for nitrogen fixation capacity. If fertilizer could be reduced in some parts of the world (high input regions), stress on regional ecosystems could be reduced. However, the possible effects of such transgene transfer to wild species are as yet not (being) studied.

• *Less harvesting from natural populations*. If product harvesting in nature for medicinal purposes, for instance could be replaced by cultivations, eventually in transgenic constructs, wild populations yield pressure may be expected to decrease. A consequence of this, which is besides the scope of this contribution, but certainly not negligible, would be that this development may change profoundly the economics of the local tribes in developing countries since it is not likely that they will be involved in the modern production methods.

• *Less pollution by mining, by production of biodegradable products, etc*. Applications will become (increasingly) available that can help to replace mineral fuels and other products, so that conventional mining can decrease, consequently, mine spoils and accompanying pollution may also go down. The same reasoning may hold for alternatives for long lasting products like plastics.

• *Genetic rescue of endangered taxa*. An extensive overview of this field has recently been presented by Benson (1999). It will be possible to genetically re-inforce threatened and depauperate taxa with known genetic deficiencies by re-installing certain alleles that have gone lost, or even may be completely new in the taxon concerned. So far we know too little of the real ecologically relevant and fitness related traits and their genetic codes, but there is no doubt that science will gain such knowledge. The expectedly high costs of such rescue operations will, however, be largely prohibitive.

• *All biota are worth preserving because of overall applicability of traits*. Since engineering is now able to transfer almost any trait from almost any taxon into almost any other, all traits present in world's biota are principally available for application. Some people argue that as a consequence, all genetic diversity, thus all biodiversity had gained conservation value (Peacock and Burdon, 1998). Some others might object that from the first data on comparative genomics, it appears that there is so much homology in the genetic code, largely irrespective of the taxon where it is found, that some more engineering will be able to manipulate a fair array of traits on the basis of the *Arabidopsis* genome alone.

Potential threats to biodiversity

In this area much of the hot debate is going on, the common denominator being the possible ecological effect of the introduction of GMO's into the environment, particularly in large scale agricultural practice. Other chapters of this volume also cover parts of this field, so I will only briefly mention some concerns.

• *Development of new weeds and invigorating of existing weeds*. There is common concern that new weeds may develop, or that existing weeds will be invigorated by genetic introgression from the crops. In this field, there are two main questions, viz., 1) what do we know about weedy traits, and 2) what makes a habitat invasible (prone to invasion) for a weed?

There is ample literature about the traits that may make a plant a weed. Baker (1974) proposed a list of particular characters, Keeler (1989) discussed and extended this, and more recently, Williamson (1996) made an evaluation, mainly on the basis of the UK flora. Although weeds in general indeed share rather a proportion of the traits identified, this holds relatively little for the UK. Predicting whether a given taxon will behave weedy or not is apparently difficult, like it is extremely difficult to predict the success of an invading species. Although the absence of most of the traits from any one taxon may indicate that the risk for sudden weed development is limited, predicting weediness is difficult, if possible at all.

The second point risen, what makes a habitat invasible, is neither easy to assess. In general, disturbance enhances invasibility. Breaks in vegetation cover, fires, and overgrazing all contribute to instability of ecosystems. Some other communities may be highly dynamic per se, disturbance being a "natural" factor: riparian ecosystems, sand dunes, waterways being well known examples. Also isolated habitats in general have high invasibility.

There is little doubt that, due to human activity in the landscapes, invasibility will increase. This will encompass not only larger surface areas, but also an increasing amount of habitats (Peacock and Burdon, 1998). Large scale logging, increase of agricultural area (particularly low-intensity systems), sub-urban developments, to name a few of the aspects, will contribute to habitat destruction and fragmentation, with inherent increase of edge effects, and will continuously increase disturbance, per area as well as in intensity (Peacock and Burdon, 1998). In principle, this is not a consequence of molecular techniques *per se*. However, it is still important to assess the effects of transgene traits in future or current constructs, taking into account the increasing invasibility.

It may be clear that any possible effect will not be exclusively connected to GMO's. Indeed, also "conventional" or, possibly better, non-transgenic races should be considered in this respect. Special emphasis should be on traits that may influence performance of wild relatives (either weeds or harmless "wild" species), changes in stress resistance may be of importance here (see, a.o., Regal, 1994). Broadly speaking, all traits that may interact with fitness parameters (including biotic as well as abiotic factors) may have an adverse impact on extant biodiversity. Traits of this kind, under study for (transgenic) transfer, are resistance genes against herbivory, pathogens, heavy metal contamination, resistance against low temperatures, high salinity levels, drought and water stress, adaptations to anaerobic root environment (see, a.o., Mannion, 1995). In general, pathogen resistance is a relatively poorly understood field, where also relations with mycorrhizal flora may be (negatively) influenced (Miller, 1993). The special concern here would be about the balance between fungi resistance and mycorrhiza. Miller (1993) suggests that constitutionally expressed Pathogen Related proteins may hamper mycorrhiza symbiosis, and consequently may reduce vitality and fitness of the plant. To my knowledge, there are still few studies available on this aspect of non-target effects. Because of the limited knowledge, virus-related questions particularly may pose problems (Hails, 2000). Some studies proved that virus resistance gene introgression may enhance fitness, but one should study the question whether wild relatives populations in fact encounter the pertaining virus infestations in nature (Bartsch et al., 1996).

• *Introgressions of transgenes into wild relatives*. Very much debate is still going on about the possible escape of transgenes into the wild flora. Of course, if there are no relatives around, this problem is non-existing, and no threat for biodiversity occurs. However, many current crop and fodder taxa do have phylogenetically close, sometimes conspecific relatives co-occurring. It has become evident that in most cases where crops and wild relatives co-occur, gene flow takes place, leading to introgression, either of the wild relative taxa into the crop (not relevant for this contribution), or from the crop into the wild relevant. Recent studies and reviews made clear that gene influx into wild species is quite common, indeed. Ellstrand et al. (1999) and, among other authors, Rieseberg et al. (1999) list such proven examples, see Table 7. In some taxa, this might indeed form a potential threat to wild relatives, or to co-occurring species in the relevant plant community. Table 7 shows which taxa were found to suffer from such genetic introgression effects to (local) extinction, or are at such risk. Note that all these reports concern traditionally bred crop races, no GMO's were involved at all.

Table 7. Introgression of crops into wild relatives. Mainly after Ellstrand et al. (1999) and Rieseberg et al. (1999). The column biodiversity risk is a qualitative summation of the extinction treat/events as given by Ellstrand et al. (1999), and the risk for biodiversity that may follow from the development of new weeds.

Crop	Introgression	Weed development	Extinction	Biodiversity risk
Triticum aestivum	Yes	No	No	No
T. turgidum	Yes	Yes	No	Yes
Oryza sativa	Yes	Yes	Yes	Yes
O. glaberrima	Yes	No	No	No
Glycine max	Yes	Yes	No	Yes
Gossypium hirsutum	Yes	No	Yes	Yes
G. barbadense	Yes	No	No	No
Sorghum bicolor	Yes	Yes	No	Yes
Pennisetum glaucum	Yes	Yes	No	Yes
Phaseolus vulgaris	Yes	Yes	No	Yes
Brassica napus	Yes	No	No	No
B. rapa	Yes	No	No	No
Helianthus annuus	Yes	Yes	No	Yes
Medicago sativa	Yes	No	Yes	Yes
Beta vulgaris	Yes	Yes	?	?
Chenopodium quinoa	Yes	?	?	?
Cucurbita pepo	Yes	?	?	?
Setaria italica	Yes	?	?	?

In hybridization and introgression, two processes may lead to negative effects on biodiversity, outbreeding depression and "hybrid vigour". Particularly small populations of wild relatives in the neighbourhood of large crop areas will be at risk of genetic swamping as an effect of outbreeding depression. There will be so much pollen from the crop around that much of the seed set in the wild relative will be hybrid ("pollen pressure" on the wild relatives). Hybrids then may be less well adapted (break down of adaptive gene complexes), and consequently, the small population will disappear, and so will the local adaptations. In fact, this process is quite similar to the introgression of common species into rare ones, discussed in the first part of this contribution. Hybrid vigour may occur if introgression affects fitness of the wild taxon positively. This is to happen if traits are introgressing that confer fitness increases, for instance any sort of stress resistance increase, compare the list of above traits.

The mechanism of pollen pressure has become clear from several recent studies (Levin et al. 1996, a review; Hokansen et al., 1997: Cucumber; Jørgensen et al., 1996: *Brassica napus* and *Brassica campestris*, depending on number of surrounding plants, 9-93% gene flow into the wild relative; den Nijs, *Lolium perenne*, see Table 6). Upscaling of the deployment of new races (of any kind) will increase this pollen pressure on crop surrounding biodiversity, and its effects.

One must expect that introgressions will take place also from future (transgenic) races, and indeed these may lead to the incorporation of new traits into the wild genome. Studies of the consecutive effects are still needed.

• *Agriculture encroachment on sensitive ecosystems*. In fact, this aspect forms part of the above, where new stress-related traits were discussed. Agricultural practice and land use may also change simply because crop races become available that are able to grow in new, not yet used environments (salt stressed, drought stressed). If so, ecosystems that are likely to be relatively

Table 8. Numbers of insect species living and at least partially depending on Scotch broom (*Cytisus scoparius,* Fabaceae). Based on data from Waloff (1968) and Ellis (unpublished data).

Part of plant	Species	Parasites and predators (including spiders)
in seeds	2	7
in pods	3	3
pods, gal inducing	2	4
leaves, herbivores	4	7
leaves, miners	2	9
borers	3	14
suckers	12	23
additionally, still to be studied:	butterfly larvae	c. 70
	flower visitors	c. 60
	various groups	c. 50
	parasitic and saprophytic fungi	c. 40
	Total number species involved	c. 320

sensitive to disturbance will become at risk of agricultural development. The effect on biota may be even larger in such systems than in the "easy growing", ecologically more relaxed sites (Peacock and Burdon, 1998).

• *Changes in ecosystems*. Changes in agricultural practices and genetic introgression may lead to cascade effects in ecosystems that, by definition, will be very difficult to predict. However, it should be expected that untraceable losses of biodiversity may be among the consequences. Such losses will not be limited to wild relatives, but very likely there will be non-target taxa among them, taxa that are part of the multi-trophic food web. I don't like to recall the Monarch butterfly story as a non-target affected taxon, there is rightly a lot of debate on that case. However, in general the situation may be much more subtle. The balance in natural herbivory resistance patterns is as yet not well known, but for sure is very complicated and delicate, indeed. It is a well established fact now that many Arthropod species are dependent on a single host plant species, tropical trees are famous for having tens of such "epiphytic" taxa. However, also in temperate regions strong data are available on similarly complex inter-dependent relationships. *Cytisus scoparius* (Scotch broom) is an example of the incredible multifacetted ecosystem relations between a single plant species and a whole spectrum of insect species and other "cryptobiota", which maintain relations with or are dependant on each other in multitrophic systems, see Table 8. Up to now, we know only very little of the effects of disturbing one or more components in complicated multitrophic chains. Being aware that we never will be able to unravel all of this, it calls for caution, if not for the precautionary principle, and we first of all should perform careful and appropriate studies.

Conclusions

From the first part of this contribution, it will be quite clear that molecular tools provide many new options to help solving conservation problems, however, in some situations there is still debate to find the equilibrium between the molecular and the taxonomic data. It is also a prerequisite that for a full and responsible valuation of the conservation interests, many ecological and demographic data are needed. Long term ecological and evolutionary perspective for biodiversity will only be regained after a careful integration of all information possibly available. Only then, we will be able to develop strategies to maintain the dynamic nature of earth's biodiversity.

It is good to note that conservation (biology) does not aim at maintaining the status quo *per se* (which might bring about the suspicion of being merely conservative), the real concern is how to achieve, as much as possible, the maintenance of the dynamics of the evolutionary process.

The second part of this chapter may indicate that there are, indeed, a lot of beneficial options connected to the implementation of the modern techniques into agricultural practice, However, there is also the obvious need for a careful, scientifically sound, and (politically and economically) unbiased analysis of the potential negative aspects. This should give the possibility for a fair weighing of the (possible) beneficial and detrimental effects of future and current practices. However, even effects of the latter are not in all cases well known.

References

Aagaard JE, Volmer SS, Sorensen FC, Strauss SH (1995) Mitochondrial DNA products among RAPD profiles are frequent and strongly differentiated between races of Douglas fir. *Mol. Ecol.* 4: 441-447

Allendorf FW, Leary RF (1986) Heterozygosity and fitness in natural populations of animals. In: Soule ME (ed.): *Conservation biology*, Sinauer, Sunderland. 57-76

Allendorf FW, Waples RS (1995) Conservation and genetics of salmonid fishes. In: JC Avise, Hamrick JL (eds.) *Conservation genetics: case histories from nature*. Columbia University Press, New York. 238-280

Ashley MV (1999) Molecular Conservation Genetics. *American Scientist* 87: 28-35

Avise JC (1992) Molecular population structure and the biogeographic history of a regional fauna: a case history with lessons for conservation biology. *Oikos* 63: 62-76

Avise JC, Nelson WS (1989) Molecular genetical relationships of the extinct Dusky Seaside Sparrow. *Science* 243: 646-648

Baker HG (1974) The evolution of weeds. *Ann. Rev. Ecol. & Systematics* 5: 1-24

Baker CS, Cipriano F, Palumbi SR (1996) Molecular genetic identification of whale and dolphin products from commercial markets in Korea and Japan. *Mol. Ecol.* 5: 671-685

Bartsch D, Schmidt M, Pohl-Orf M, Haag C, Schuphan I (1996) Competitiveness of transgenic sugar beet resistant to beet necrotic yellow vein virus and potential impact on wild beet population. *Mol. Ecol.* 5: 199-205

Bininda-Emonds ORP, Vázquez DP, Manne LL (2000) The calculus of biodiversity: integrating phylogeny and conservation. *Trends Ecology & Evolution* 15: 92-95

Benson EE(ed.) (1999) *Plant conservation biotechnology*. Taylor & Francis, London. 309 pp

Bleeker W, Huthmann M, Hurka H (1999) Evolution of hybrid taxa in Nasturtium R.Br. (Brassicaceae). *Folia Geobotanica* 34: 421-433

Boer ER de, van Leeuwen NC, den Nijs JCM, Ellers J, Willems JH (1995) De toestand van Orchis-soorten in Nederland, genetisch bezien. *Rapport Universiteit van Amsterdam/Universiteit Utrecht.* 42 pp. & Appendices

Bowen BW, Nelson WS, Avise JC (1993) A molecular phylogeny for marine turtles: trait mapping, rate assessment, and conservation relevance. *Proceedings of the National Academy of Sciences of the USA* 90: 5574-5577

Brownlow CA (1996) Molecular taxonomy and the conservation of Red Wolf and other endangered carnivores. *Cons. Biol.* 10: 390-396

Burke JM, Carney SE, Arnold ML (1998) Hybrid fitness in the Louisiana irises: evidence from experimental analyses. *Evolution* 52: 1304-1310

Demesure B, Comps B, Petit RJ (1998): Chloroplast DNA phylogeography of the common beech (Fagus sylvatica L.) in Europe. *Evolution* 50: 2515-2520

Dumolin-Lapègue S, Demesure S, Fineschi V, Corre LE, Petit J (1997) Phylogeographic structure of white oaks throughout the European continent. *Genetics* 146: 1475-1487

Ellstrand NC (1992) Gene flow by pollen: implications for plant conservation genetics. *Oikos* 63: 77-86

Ellstrand NC, Elam D (1993) Population genetic consequences of small population size: Implications for plant conservation. *Ann. Rev. Ecol. Systematics* 24: 217-242

Ellstrand NC, Prentice HC, Hancock JF (1999): Gene flow and introgression from domesticated plants into their wild relatives. *Ann. Rev. Ecol. Systematics* 30: 539-563

Francisco-Ortega J, Jansen RK, Mason-Gamer RJ, Wallace RS (1996) Application of chloroplast DNA restriction site studies for conservation genetics. In: T.B. Smith, Wayne RK (eds.): *Molecular genetic approaches in conservation.* Oxford UP, Oxford. 183-201

Goldstein PZ, DeSalle R, Amato G, Vogler AP (2000) Conservation genetics at the species boundary. *Cons. Biol.* 14: 120-131

Gonzalez-Beinto ME, Martin C, Iriondo JM, Perez C (1999) Conservation of the rare and endangered plants endemic to Spain. In: Benson EE (ed.): *Plant conservation biotechnology.* Taylor & Francis, London. 251-264.

Haig SM (1998) Molecular contribution to conservation. *Ecology* 79: 413-425

Hails RS (2000) Genetically modified plants – the debate continues. *Trends Ecology & Evolution* 15: 14-18

Hamrick JL, Godt MJ (1989) Allozyme diversity in plants. In: Brown AHD, Clegg MT, Kahler AL, Weir BS (eds.): *Plant population genetics, breeding and genetic resources.* Sinauer, Sunderland. 43-63

Hamrick JL, Godt MJ (1996) Conservation genetics of endemic plants species. In: Avise JC, Hamrick JL (eds.): *Conservation genetics: case histories from nature.* Chapman & Hall, London. 281-304

Hamrick JL, Godt MJW, Murawski DA, Loveless MD (1991). Correlations between species traits and allozyme diversity: Implications for conservation biology. In: Falk DA, Holsinger KE (eds.): *Genetic and Conservation of Rare Plants.* Oxford. 75-86

Harris SA (1999) Molecular approaches to assessing plant diversity. In: Benson EE (ed.): *Plant conservation biotechnology.* Taylor & Francis, London. 11-24

Hokansen SC, Grumet R, Hancock JF (1997) Effect of border rows and trap/donor ratios on pollen-mediated gene movement. *Ecological Applications* 7: 1075-1081

Jørgensen RB, Andersen B, Landbo L, Mikkelsen TR (1996) Spontaneous hybridization between oilseed rape (Brassica napus) and weedy relatives. *Acta Horticulturae* 407: 193-200

Karp A, Edwards KJ, Bruford M, Funk S, Vosman B, Morgante M, Seberg O, Kremer A, Boursot P, Arctander P et al. (1997) Molecular technologies for biodiversity evaluation: opportunities and challenges. *Nature Biotechnology* 15: 625-628

Keeler KH (1989) Can genetically engineered crops become weeds? *Biotechnology* 7: 1134-1139

King RA, Ferris C (1998) Chloroplast DNA phylogeography of Alnus glutinosa (L.) Gaertn. *Mol. Ecol.* 7: 1151-1163

Klinger T, Ellstrand NC (1999) Transgene movement via gene flow: recommendations for improved biosafety assessement. In: K Ammann, Jacot Y, Simonsen V, Kjellsson G (eds.): *Methods for risk assessment of transgenic plants. III. Ecological risks and prospects of transgenic plants.* Birkhäuser, Basel. 129-140

Leberg PL (1996) Applications of allozyme electrophoresis in conservation biology. In: TB Smith, Wayne RK (eds.): *Molecular genetic approaches in conservation.* Oxford UP, Oxford. 87-103

Legge JT, Roush R, DeSalle R, Vogler AP, May B (1995) Genetic criteria for establishing Evolutionarily Significant Units in Cryan's Buckmoth. *Cons. Biol.* 10: 85-98

Levin DA, Francisco-Ortega J, Jansen R (1996) Hybridization and the extinction of rare plant species. *Cons. Biol.* 10: 10-16

Mace GM, Smith TB, Bruford MW, Wayne RK (1996) An overview of the issues. In: Smith TB, Wayne RK (eds.): *Molecular genetic approaches in conservation.* Oxford UP, Oxford. 3-21

Mandal BB (1999): Conservation biotechnology of endemic and other economically important plant species of India. In:.Smith TB, Wayne RK (eds.): *Molecular genetic approaches in conservation.* Oxford UP, Oxford. 211-225

Mannion AM (1995) Agriculture, environment and biotechnology. *Agriculture, Ecosystems and Environment* 53: 31-45

Miller RM (1993) Nontarget and ecological effects of transgenically altered disease resistance in crops – possible effects on the mycorrhizal symbiosis. *Mol. Ecol.* 2: 327-335

Möller M, Cronk QCB (1997) Phylogeny and disjunct distribution: evolution of Saintpaulia (Gesneriaceae). *Proc. Royal Soc. London B* 264: 1827-1836

Moritz C (1994) Defining evolutionarily significant units for conservation. *Trends Ecology & Evolution* 9: 373-375

Moritz C (1995) Uses of molecular phylogenies for conservation. *Philosophical Transactions Royal Society of London B* 349: 113-118

Newton AC, Allnutt TR, Gillies ACM, Lowe AJ, Ennos RA (1999) Molecular phylogeography, intraspecific variation and the conservation of tree species. *Trends in Ecology & Evolution* 14: 140-145

O'Brien SJ (1994) A role for molecular genetics in biological conservation. *Proc. Natl. Acad. Sci. USA* 91: 5748-5755

Oostermeijer JGB (1996) Population viability of the rare Gentiana pneumonanthe, the relative importance of demography, genetics, and reproductive biology. *PhD Thesis*, University of Amsterdam

Oostermeijer JGB (2000) Is genetic variation important for the viability of wild plant populations? *Schriftenreihe für Vegetationskunde* 32: 23-30

Oostermeijer JGB, van Eijck MW, van Leeuwen NC, den Nijs JCM (1995) Analysis of the relationship between allozyme heterozygosity and fitness in the rare Gentiana pneumonanthe L. *Journal of Evolutionary Biology* 8: 739-759

Oostermeijer JGB, Luijten S, Kwak M, Boerrigter E, den Nijs H (1998) Zeldzame planten in het nauw: problemen van kleine populaties. *De Levende Natuur* 99:134-141

Parker KM, Sheffer RJ, Hedrick PW (1999) Molecular variation and evolutionarily significant units in the endangered Gila Topminnow. *Cons. Biol.* 13: 108-116

Pavlik BM, Nickrent DI, Howald AM (1993) The recovery of an endangered plant. I. Creating a new population of Amsinckia grandiflora. *Cons. Biol.* 7: 510-526

Peacock WJ, Burdon JJ (1998) Biodiversity, biotechnology and agriculture. *Pure and Applied Chemistry* 70: 2101-2106

Pence VC (1999) The application of biotechnology for the conservation of endangered species. In: Benson EE (ed.): *Plant conservation biotechnology*. Taylor & Francis, London. 227-250

Regal PJ (1994) Scientific principles for ecologically based risk assessment of transgenic organisms. *Mol. Ecol.* 3: 5-13

Rieseberg LH, Min Ju K, Seiler GJ (1999) Introgression between the cultivated sunflower and a sympatric wild relative, Helianthus petiolaris (Asteraceae). *Int. J. Plant Sc.* 160: 102-108

Rosetto M, Lucarotti F, Hopper SD, Dixon KW (1997) DNA fingerprinting of Eucalyptus graniticola: a critically endangered relict species or a rare hybrid? *Heredity* 79: 310-318

Rufener al Mazyad. P, Ammann K (1999) The Medicago falcata/sativa complex, crop-wild relative introgression in Switzerland. In: van Raamsdonk LWD, den Nijs JCM (eds): *Plant evolution in man-made habitats*, Hugo de Vries Laboratory, Amsterdam. 271-286

Ryder OA (1986) Species conservation and systematics: the dilemma of subsecies. *Cons. Biol.y* 1: 9-10.

Ryman N, Utter F, Hinder K (1995) Introgression, supportive breeding, and genetic conservation. In: Ballou JD, Gilpin M, Foose TJ (eds.): *Population management for survival and recovery.* Columbia University Press, New York, 341-365

Sewell MM, Parks CR, Chase NW (1996) Intraspecific chloroplast DNA variation and biogeography of north American Liriodendron L. (Magnoliaceae). *Evolution* 50: 1147-1154

Sinclair WT, Morman JD, Ennos RA (1999) The postglacial history of Scots pine (*Pinus sylvestris* L.) in western Europe: evidence from mitochondrial DNA variation. *Mol. Ecol.* 8: 83-88

Soltis PS, Gitzendanner MA (1999) Molecular systematics and the conservation of rare species. *Cons. Biol.* 13: 471-483

Smith TB, Wayne RK (eds.) (1996) *Molecular genetic approaches in conservation*. Oxford UP, Oxford. 483 pp.

Taberlet P (1996) The use of mitochondrial DNA control region sequencing in conservation genetics. In: TB Smith, Wayne RK (eds.), *Molecular genetic approaches in conservation.* Oxford UP, Oxford. 125-142

Tomaru N, Mitsutsuji T, Takahashi M, Tsumura Y, Uchida K, Ohba K (1997) Genetic diversity in Fagus crenata (Japanese beech): influence of the distributional shift during the late-Quaternary. *Heredity* 78: 241-251

Urbanska KM (1994) Ecological restoration above the timberline - demographic monitoring of whole trial plots in the Swiss alps. *Botanica Helvetica* 104: 141-156

Urbanska KM (1995) Ecological restoration above timberline and its demographic assessment. In: Urbanska KM, Grodzinska K (eds.): *Restoration ecology in Europe*. Geobotanical Institute SFIT Zürich, 15-36

Waloff N (1968) Studies on the insect fauna on Scotch broom *Sarothamnus scoparius* (L.) Wimmer. *Advances in ecological Research* 5: 87-208

Warwick SI, Small E (1999) Invasive plant species: evolutionary risk from transgenic crops. In: van Raamsdonk LWD, den Nijs JCM (eds.): *Plant evolution in man-made habitats*, Hugo de Vries Laboratory, Amsterdam. 235-256

Williams SL, Davis CA (1996) Population genetics analyses of transplanted Eelgrass (Zostera marina) beds reveal reduced genetic diversity in Southern California. *Restoration Ecology* 4: 163-180

Williamson M (1996) Community response to transgenic plant release: prediction from British experience of invasive plants and feral crop plants. *Mol. Ecol.* 3: 75-79

Young A, Boyle T, Brown T (1996) The population genetic consequences of habitat fragmentation for plants. *Trends Ecology & Evolution* 11: 413-418

Methods for Risk Assessment of Transgenic Plants
IV. Biodiversity and Biotechnology
K. Ammann, Y. Jacot and R. Braun (eds)
© 2003 Birkhäuser Verlag Basel/Switzerland

Conservation strategies, plant breeding, wild species and landraces

Vernon Heywood

Centre for Plant Diversity and Systematics, School of Plant Sciences, The University of Reading, Reading RG6 6AS, UK

"*Biological diversity is the ultimate source of both culture and agriculture*" (Busch et al. 1989).

Introduction: biodiversity and agrobiodiversity

The problems of world poverty will not be solved by biotechnology alone, nor by traditional plant breeding or improved agronomic practices but by a combination of all of these and other approaches. We need therefore to adopt a broad approach to our considerations of the links between biodiversity, agrobiodiversity and biotechnology.

As noted in the Brazilian Government proposal to SBSTTA of the CBD, Second Meeting, Montreal 2-6 September 1996, 'Biodiversity is a non-detachable part of the concept of sustainability. ... biodiversity is essential for agricultural production, as agriculture should be for biodiversity conservation.'

We should remember in our debates on biotechnology, that agriculture and forestry is a complex mix that involves a whole series of elements of agrobiodiversity – in terms of the agroecosystems, genetics and the cultivated units, as well as the diversity of socio-cultural systems and interactions involved (Table 1).

Agricultural biodiversity, also known as agrobiodiversity, is as difficult to define precisely as biodiversity itself. As a general definition it can be regarded as biodiversity in an agricultural context and described as the variety and variability amongst living organisms (of animals, plants and micro-organisms) that are important to food and agriculture in the broad sense and associated with cultivating crops and rearing animals and the ecological complexes of which they form a part. It is not just a subset of biodiversity but an extension of it in that it embraces units (such as landraces, cultivars, pure lines, and strains) and habitats (agroecosystems such as farmers' fields) that are not normally considered or even accepted as properly part of biological diversity (Heywood 1999a). In fact some authors exclude artificial diversity, such as species introduced into an area (thus including by implication agricultural crops) from the concept of biodiversity 'because it cannot fulfil the full range of societal values that native biodiversity does' (Angermeier 1994).

Agrobiodiversity includes all those wild species and the landraces, crop varieties, animal breeds and races, and microorganism strains derived from them, that we use directly or indirectly for food and agriculture, both as human nutrition and as feed (including grazing) for domesticated and semi-domesticated animals, as well as industrial and fuel crops, together with the whole range of environments in which agriculture is practised, not just the crop lands or fields themselves. It also includes many habitats and species outside of farming systems that benefit agriculture and enhance ecosystem functions. The Global Plan of Action (FAO 1996) notes, for example, that 'Natural ecosystems hold important plant genetic resources for food and agriculture, including endemic and threatened wild crop relatives and wild plants for food production'.

Table 1. The components of agrobiodiversity (from Heywood 1999a)

Agroecological diversity	**Organismal diversity**	**Genetic diversity**
biomes	kingdoms	gene pools
agroecological zones	phyla	populations
agroecosystems	families	individuals
polycultures	genera	genotypes
monocultures	species	genes
mixed systems	subspecies	nucleotides
rangelands	varieties	
pastures	cultivar groups	
fallows	cultivars	
agroforestry systems	landraces	
agrosylvicultural	breeds	
sylvopastoral	strains	
agrosylvopastoral	pure lines	
home gardens		
forest ecosystems		
managed forests		
plantation forests		
seed forests		
fisheries		
freshwater systems		
marine systems		
habitats		
field		
plot		
crop		
Socio-cultural diversity: human interactions with the above at all levels		

Agrobiodiversity, like biodiversity proper, can be considered at three main levels – those of ecological diversity, organismal diversity and genetic diversity (Heywood 1995), each of which forms a hierarchy of elements (Table 1). A more extensive hierarchy of agroecosystems has been given by Conway (1987, 1997) and reproduced as Fig. 1.

Socio-cultural diversity: human interactions with the above at all levels

The recognition of agrobiodiversity as a concept and as an issue, is a major conceptual breakthrough. It is not only supported by the Global Plan of Action but, as discussed below, by the Convention on Biological Diversity. It represents 'the emergence of a new paradigm for agriculture that embraces not just the most technologically advanced and efficient farming and production systems, dependant on highly bred or engineered crops and animal breeds, such as GMOs, with an emphasis on uniformity and standardization, and based on a very restricted set of species, but also recognizes that the great diversity of traditional farming systems and practices in many cultures in different parts of the world and the thousands of species that are locally cultivated or semi-domesticated in home gardens or other polycultures, or harvested from the wild in nearby habitats make a major and essential contribution to food security for hundreds of millions of people across the globe and make a major contribution to

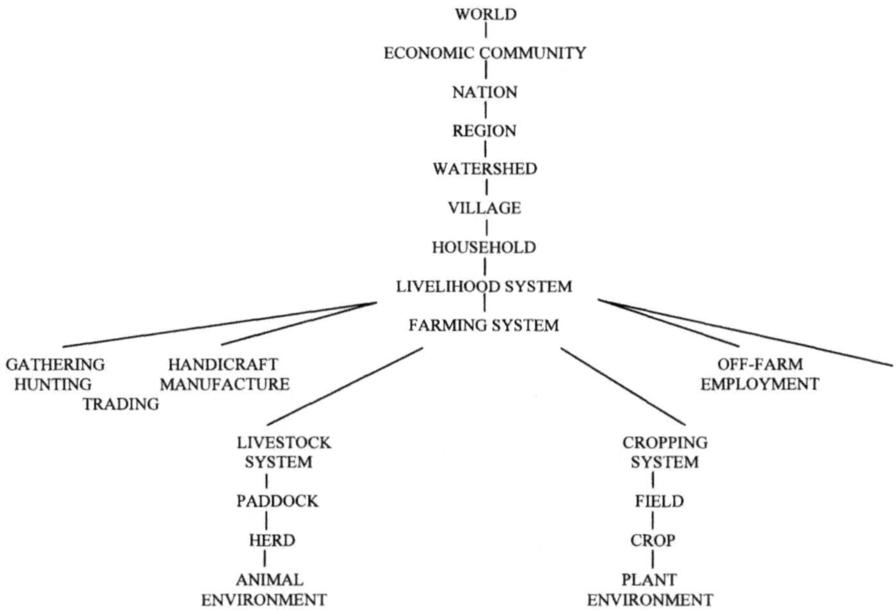

Figure 1. The hierarchy of agroecosystems (from Conway, 1997).

sustainable agriculture' (Heywood 1999a).

Although sustainable agriculture, cannot be rigidly defined, it is widely interpreted as consisting of practices that are ecologically sound, socially responsible and economically viable (Thrupp 1996). This has led many to look at traditional forms of agriculture where diversity plays a major role. It is estimated that about 60% of the world's agriculture consists of traditional subsistence farming systems in which there is both a high diversity of crops and species grown and in the ways in which they are grown, such as polycropping and intercropping, that leads to the maintenance of greater or lesser amounts of variation within the crops.

As McCalla and Brown (2000) note, at least 90% of the growth in population will occur in countries located between the tropics of Cancer and Capricorn and since about 90% of the world's food consumption takes place where the food is produced, it follows that 90% of the increased food production we need must be grown in farming systems prevalent in tropical countries. Thus we have to consider the impacts on biotechnology not only on large-scale cultivation of staples but medium and small-scale operations and the 1.4 billion farm families already mentioned who rely in varying degrees on small-scale cultivation of staples, and various forms of traditional agriculture, including diverse agroforestry polycultures (e.g. home gardens), semi-domesticated species and wild harvesting of fruits, fibres, medicinals and so on, and on the natural and semi-natural ecosystems that border or are adjacent to the cultivated fields. These humid, subhumid, and semi-arid systems are highly complex with heterogeneous mixtures of annuals, perennials, trees and livestock. Our knowledge if these is poor: we need is to assess, inventory, monitor and try and understand this diversity of species and varieties grown, as well as the different approaches in traditional farming systems and in various forms of wild harvesting and extractivism. We must also capture the traditional knowledge on agriculture, cultural practices and uses which is rapidly disappearing as modern cropping systems take over and the older generations die out.

According to Altieri (1999), it has been estimated that more than three million hectares survive under traditional agriculture as raised fields, terraces, swidden fallows, polycultures, home gardens and other agroforestry systems. Although these seldom produce marketable surpluses or have the capacity to do so, they do make a major contribution to farm household food security (Heywood 1999b) and traditional cropping systems are estimated to provide as much as 20% of the world's food supply.

Most of the discussions on biotechnology and GMOs concern a small number of staple food crops or major industrial crops but we need to remember that globally agriculture involves an enormous diversity of species. Worldwide many thousands of plant species are used in human activities and can be considered resources. About 6-7,000 have at one time or other been cultivated but of the 250–300 000 known higher plant species, only a few hundred species have been fully domesticated and enter world trade.

While only 103 species of food plants contribute 90% of national per capita supplies (Prescott-Allen and Prescott-Allen 1990), and 20-30 of these are regarded as the staples that supply most of human nutrition needs, thousands of species are grown locally that are scarcely or only partially domesticated and many thousands more are gathered from the wild. Not surprisingly, most of the partially domesticated or wild-collected species are found in the tropics. Altieri (1999) estimates that throughout the Third World, more than 3,000 native grains, roots, fruits and other food plants can still be found. In addition, more than a thousand plant species are used in human activities in Mediterranean and temperate regions of the world. An estimate of not just food plants but of species belonging to all commodity groups, can be obtained by considering the Plant Resources Project of South-East Asia (PROSEA) that records approximately 6000 species in its Basic List of species (some of them exotic) used by humankind in that area (Jansen and al. 1993). If we assume similar levels in other tropical regions, we can extrapolate to a figure of 18–25 000 species used by humans for the tropics as a whole.

These figures exclude most of the 25 000 species that are estimated to have been used or are still in use as herbal medicines in various parts of the world, especially China, tropical Asia, the Indian subcontinent, Africa and Central and South America, and the many thousands of species are grown as ornamentals in parks and in public and private gardens and in the horticultural trade.

It is only recently that a broadening of the basis of plant genetic resources and their conservation has begun to be accepted (cf. Heywood 1993), largely as a result of the Global Plan of Action as discussed below.

The context: the CBD and the Global Plan of Action

The global biodiversity context within which this workshop was held is a dual one – firstly, the Convention on Biological Diversity, signed in 1992 and subsequently ratified by 182 countries, a landmark development that put biological diversity onto the political agenda, and secondly, as already mentioned, the endorsement by governments of the 'Global Plan of Action (GPA) for the Conservation and Sustainable Use of Plant Genetic Resources for Food and Agriculture' at the International Technical Conference on Plant Genetic Resources held in Leipzig, 17–23 June 1996, another key development.

Bringing the provisions of the Convention on Biological Diversity into effect has led to a considerable number of activities and initiatives as governments attempt to get to grips with the problems of implementing the requirements of the various articles of what is in effect only an outline convention. Some, such as Raven (1998) have criticized the deliberations of the Conference of the Parties (COP) of the Convention and its Subsidiary Body for Scientific

Technical and Technological Advice (SBSTTA) for spending too much time on issues such as biotechnology, biosafety and intellectual property rights, while issues such as the preservation of species has been ignored. Others such as Seyani (1998) regard the Convention as providing incentives for countries to conserve and sustainably use their own biodiversity and ensure that the benefits derived from it by third parties are equitably shared. What is certain is that a great deal of controversy has been generated by issues such as access to genetic resources, farmers' rights and technology transfer, reflecting different attitudes and perceptions between developing and developed countries. These are, of course, important issues that demand attention serious debate, and while some of us who are alarmed at the growing rate of biodiversity loss and habitat destruction are naturally impatient and would like to see more urgent steps being taken to stem the losses, progress in reconciling the different positions is being made and at the same time many countries are taking positive steps to put in place mechanisms for the conservation and sustainable use of their biodiversity. Despite the undoubted difficulties of implementing the CBD, substantial progress has been made although major problems of funding hinder many of the necessary actions.

A significant development that followed from these two was the convergence of interest between bodies such as FAO and IPGRI and conservation and development organizations and agencies such UNESCO-MAB, IUCN, WWF and ITDG. On the one hand, the CBD recognized agricultural biodiversity as a focal area in view of its social and economic relevance and the prospects offered by sustainable agriculture for reducing the negative impacts of biological diversity, enhancing the value of biological diversity and linking conservation efforts with social and economic benefits (Decision III/111 Conservation and Sustainable Use of Agricultural Biological Diversity of the Conference of the Parties to the Convention on Biological Diversity). On the other hand, it is recognized that the Global Plan of Action covers a number of multidisciplinary areas such as *in situ* conservation of wild plants and crop relatives in natural ecosystems that extend the traditional activities of sustainable agriculture and plant genetic resource conservation, and that successful implementation will require the development of new partnerships with a range of intergovernmental and non-governmental organizations as well as with indigenous and local communities.

Biotechnology and traditional small-scale agriculture

It is widely agreed that there must be a considerable acceleration if efforts to increase the productivity of small-scale local agriculture. As Brush (1999) indicates, 'Perhaps the most important strategy for increasing the value of local crops is to use them as the basis for crop improvement programs, especially with the participation of farmers who will use the results'. This is known as 'participatory plant breeding' and can involve development of local crops through techniques such as mass selection and may be especially suitable for marginal environments where conventional breeding has limited success.

We must ask, however, how far can biotechnology assist local farmers enhance the productivity of their local crop landraces and semi-domesticates on which they are dependant to a greater or lesser degree, as well as the staples or cash crops which provide the larger part of their basic nutrition or income. As the CSD Intersessional ad hoc Working Group on Integrated Planning and Management of Land Resources and Sustainable Agriculture and Rural Development (28 February – 3 March 2000) notes, this has to be done in a way that respects traditional farming practices and the rights of indigenous and local communities.

It is highly unlikely that much effort will be put into the improvement of local crops by the mainly private sector agricultural biotechnology companies and so far only small amounts of transgenic seed have been grown in the poorer of the developing countries. Pinstrup-Andersen and Cohen (2000) believe that the direct transfers of the products of agricultural biotechnology research to developing countries will not work in most cases. What might be more appropriate is use a combination of conventional plant breeding and tissue conventional and uncontroversial biotechnology techniques such as tissue and cell culture to develop alternative forms of weed resistance. An example is the leafier forms of rice that deny weeds sunlight, developed by the West African Rice Development Association (WARDA), as a result of participatory varietal selection (WARDA 1999).

Molecular genetic techniques are being used for studying the distribution and extent of genetic variation within and between species and in the characterization of accessions, and increasingly laboratories with the capacity for undertaking various types of molecular analysis are being established in developing countries (Ayad et al. 1997). For example, the use of molecular markers have been applied in studying the genetic diversity of rattan species, a major resource in Southeast Asia that is at serious risk of depletion through forest loss and uncontrolled exploitation (Changtragoon et al. 1997).

Another approach, that is dependant on recombinant DNA experiments, is the plant breeding programme being undertaken in India in anticipation of climatic change and sea level rise, to breed salt-tolerant varieties of rice and other crops in coastal areas, to prepare for seawater intrusion into farmland as a result of an eventual rise in sea water level, using a mangrove species as the salt tolerance donor (Swaminathan 2000).

More controversial, as we have seen, are recent attempts to address problems of micronutrient malnutrition in developing countries by introduction of appropriate genes from other species into staples.

There is, as we have seen, considerable potential, although so far of limited value in practice, for the application of biotechnology in helping solve the problems of traditional agriculture in developing countries. The situation has been well summarized by Pinstrup-Andersen and Cohen (2000): '...the biggest risk of modern biotechnology for developing countries is that technological development will bypass poor farmers and poor consumers because of a lack of enlightened adaptation. It is not that biotechnology is irrelevant, but that research needs to focus on the problems of small farmers and poor consumers in developing countries. Private sector research is unlikely to take on such a focus, given the lack of future profits'. There is a need for a stronger public sector role, including the NARs and IARCs in partnership, so as to avoid a situation developing whereby cutting edge science becomes orientated exclusively toward industrial countries and large-scale farming. Agricultural biotechnology can make a contribution to food security in developing countries, provided it focuses on the need of poor farmers and works in consultation and partnership with them. The conference on biotechnology and its potential impact on agriculture in developing countries, held at the World Bank in Washington on October 21–22, 1999, sponsored by the CGIAR and the US National Academy of Sciences, and published under the title 'Biotechnology for the Poor' (Persley and Langton 2000), addresses these issues in detail.

It has been estimated that approximately 1.4 billion people live in farm families that are largely self-reliant and self-provisioning for their seeds and other planting materials. The maintenance of these seeds may be considered by them as advantageous or they may have no alternative sources of seed available that meet their needs or that they can afford (Fowler et al. 1999). It is clear that the diversity of landraces in farmer's fields will continue to exist for some considerable time at least in some parts of the world.

But even though it is accepted that the production and maintenance of such local cultivars on farms have a significant role in maintaining intra-specific diversity and plays an important part in farm incomes and in human nutrition and other needs, it has to be recognized our knowledge of the both the mechanisms and the economics involved is quite limited.

Although numerous studies have been made of such systems, that involve traditional landraces and tens of thousands of semi- or sub-domesticated species, we are far from having an overview of their contribution to global human nutrition. In fact, it is not known what are the relative contributions made by the different types of crop (commercial cultivars of major crops, minor crops, landraces, primitive cultivars, semi-domesticates etc.) to global agriculture or to human nutrition.

Landraces

Much of the focus on genetic resources in traditional agriculture is on landraces. Landraces, primitive cultivars, or farmers' varieties are complex units and difficult to define precisely and even more difficult to conserve. They are the products of breeding or selection carried out by farmers, either deliberately or not, over many generations. In the words of Harlan (1975), they 'are recognizable morphologically; farmers have names for them, and different landraces are understood to differ in adaptation to soil type, time of seeding, date of maturity, height, nutritive value, use and other properties. Most important, they are highly genetically diverse. They are balanced populations – variable, in equilibrium with both environment and pathogens, and genetically dynamic.' They may be grown as mixtures of species or as diverse populations of a single species (Hawkes 1983).

An essential feature of landraces is that they have become adapted to the conditions under which they are cultivated and thus may be regarded as the equivalent to ecotypes of wild species (Holden et al. 1993); they were indeed recognized as part of the genecological system by Gregor (see Gregor and Horne 1935) who termed them 'agroecotypes'. Thus while ecotypes are the result of natural selection acting on particular genes or gene combinations that are most adapted the particular factors of the local environment (Davis and Heywood 1963), in the case of adaptive change in cultivated species, we must add additional selection pressures caused by different cultivation and harvesting practices, whether applied consciously or unconsciously by the farmer. In other words landraces are the result of both natural and farmer selection. Another major difference from ecotypes is of course that landraces can arise in a range of different habitats, often far from the areas to which the parental species is native. It is in this sense that agriculture may be said to enhance biodiversity although not, of course, of natural plant or animal populations.

Landraces, as we have seen, are but one of the components of genetic diversity of concern to agriculture and plant breeding but because of their genetic diversity are the focus of most agricultural germplasm conservation efforts. The process of plant breeding in response to changing conditions, disease, cultivation practices, climatic change, human requirements, is dynamic and endless.

Conservation strategies for biodiversity and agrobiodiversity

The focus of nature conservation has over the past few decades been a dual one – on the development of protected areas such as National Parks and Biosphere Reserves, i.e. ecosystem-based *in situ* conservation, on the one hand, and on endangered species, Red Lists, species recovery plans, etc., i.e. a species-based conservation on the other. There has been a tendency to polarize this situation, and as Soulé and Mills (1992) put it, dichotomize nature into species and ecosystems, and conservation genetics, because it is a species or population based approach has in turn has been deemed guilty by association with species chauvinists. As Lawton and Jones (1993) have noted, for almost three decades ecosystem and population ecology have ploughed their own independent furrows and developed their own paradigms, approaches and questions.

Because of the enormity (and, some would say futility) of the task of species conservation, many conservationists have criticised a species-based approach which is essentially a crisis management approach, as opposed to a habitat, ecosystem or landscape approach. This anti-species backlash has manifested itself recently in the Convention on Biological Diversity through its advocacy of 'the ecosystem approach' although when analyzed this embraces both species and ecosystems as they are mutually interdependent. As Soulé and Mill (1992), observe, a pure ecosystem approach is as absurd as a pure species approach.

Recently it has been increasingly recognized that because of the limitations of species-based and ecosystem-based approaches and taking into account the dynamics of species and ecosystems and the effects of human action, integrative/holistic/complementary methods for establishing conservation priorities should be adopted. Essentially this recognizes that one should adopt whatever scientific and social techniques or approaches, such as *in situ*, *ex situ*, *inter situ*, reintroduction, population reinforcement, that are judged appropriate to a particular case and circumstances (Falk 1990).

Conservation has only become an explicitly recognized concern in agriculture fairly recently, although the practice of local farmers in conserving genetic diversity through maintaining heterogeneous landraces in cultivation, has been happening for centuries. This was eventually followed by a phase of selection of cultivars in the 19th century, that led to the supplanting of landraces and the beginnings of genetic erosion, a process that was intensified as modern high-yielding cultivars (HYVs) were developed. The roots of the modern movement for the conservation of genetic diversity can be found in the 1930s and 1940s as breeders in many countries, following the lead given by Vavilov[1], began to assemble collections of material that could be of potential use in breeding programmes (Holden et al. 1993). This was done through gathering and exchanging material between breeders and by commissioning or participating in collecting expeditions, especially in areas where the crops in which they were interested showed high diversity. The collections were generally not well maintained as their role was seen more as an adjunct to plant breeding than as resources to be conserved[2].

As awareness developed of the disappearance of landraces and primitive varieties from farmers fields, the modern official genetic resources movement began, with the genetic resources collection and conservation campaign that began in the 1960's, the formation of the FAO Genetic Crops Unit in 1965, and later, within the framework of the CGIAR, the founding of IBPGR (later to become IPGRI) in 1974. The main basis of genetic conservation of agriculturally important germplasm was, and still is, storage of seed in secure conditions in gene banks, i.e. *ex situ*. The gene banks collections that developed house mainly accessions of landraces and cultivars with wild species typically account for less than 2% of the accessions in crop germplasm collections (Plucknett et al. 1987).

Although the importance natural ecosystems and the wild species they house was recognized by Vavilov and others, little more than lip service has been paid to this by the agricultural genetic resources sector, largely because of their mandate to focus on the staple crops and a lack of resources to undertake any significant effort on *in situ* conservation.

The nature conservation movement itself has largely ignored genetics until recently. Indeed, as Soulé and Mills (1992) commented, 'Until the middle 1970s, most of the people in charge of conservation ignored genetics, and most of the people in charge of genetics ignored conservation'. Even today, when so many ecosystems (and component species populations)

---

[1]Vavilov himself, through his work at what became the All-Union Institute of Plant Industry in Leningrad and the more than 400 hundred research institutes and experimental stations under his control, with a staff of some 20 000 under his orders, amassed more than 250 000 accessions of germplasm through an extensive series of collecting exepeditions (Hawkes 1990).

[2]An alternative chronology of *ex situ* conservation is given by Cohen 1992) who recognizes four eras:

1. Era of plant exploration and introduction: 1850–1950
2. Era of conservation: mid-1950s–1980s
3. Era of regeneration and new international linkage: 1980s
4. Era of more efficient utilization: 2010 and beyond

have been subjected to fragmentation, the loss of genetic diversity of the species' populations that occur in habitat fragments is not only of interest on theoretical grounds but is a matter of serious concern in respect to the maintenance of genetic resources of species that have an agricultural or other economic values. Yet, such 'genetic erosion' tends to be overlooked in discussions of fragmentation and conservation (Turner 1996), perhaps because many fragmented species populations are more likely to become extinct through stochastic factors rather than loss of genetic variability; if, however, conservation of genetic variation is a focus of concern, as for example, in crop relatives then such arguments lose their force.

Although the World Conservation Strategy published in 1980 recognized the importance of agriculture and included a programme for the conservation of zones rich in genetic resources, subsequently mainstream conservation gave genetic resources little priority, preferring to concentrate on he identification and protection of charismatic or flagship species, or any kind of endangered species, and on setting up and managing Protected Area Systems, with a particular emphasis on biodiversity-rich tropical forests. Certainly, lip service was paid to the conservation of wild species used by humans in campaigns such as 'Saving the Plants that Save Us' of WWF, although agriculture was generally viewed as detrimental to the conservation of biodiversity; and occasionally attention was drawn to importance of wild species in tropical forests in terms of their value to traditional extractivists, as a justification for their conservation rather than conversion to other uses. But with the exception of the pioneering work of Altieri and others on agroecology and the importance of traditional farming systems not only in terms of cultivated fields but the natural ecosystems that surround them, and Nabham on local seed savers, conservationists and conservation biologists alike have contrived to avoid these issues.

Vandermeer and Perfecto (1997) addressed the issue of the agroecosystem and the need for viewing it under the conservation biologists lens. They observe that the failure of conservation biologists to show much interest in agroecosystems, simply because they are already tainted, is matched by the lack of interest from agroecologists in the biodiversity found in traditional agroecosystems since it has no obvious connection to production. As we are increasingly concerned with introducing forms of sustainable agriculture that will reduce the adverse effects of agriculture on biodiversity, it is vital that the conservation biology community changes its attitude and starts considering agroecosystems as legitimate areas for study and begin asking the same questions that they do about so-called 'natural ecosystems'.

Conservation as management for change

Conservation of biodiversity has undergone a major phase of reappraisal in recent years. One of the greatest changes in our perception of conservation stems from an increasing realization of the amount of dynamism that exists in natural systems. This is reflected in attitudes to *in situ* conservation, which is now recognized as covering a series of different situations (Figure 2). *In situ* conservation, whether it be of species' populations or 'ecosystems' (or, as we shall see of landraces on-farm) is a means of ensuring that evolutionary change continues.

Conservation is essentially management for change. Unlike *ex situ* preservation, it is a means of allowing the dynamic processes in ecosystems and populations to continue to take place. In fact, one of the main justifications for *in situ* conservation (and, at the same time, a failing of *ex situ* conservation) is that it allows evolutionary change to continue in the component species and populations but in addition to endogenous evolutionary (and ecological) change, a whole series of exogenous factors are also involved. We have moved on from the earlier 'laissez faire' ethic that dominated conservation philosophy in earlier decades towards acceptance of a much more interventionist approach on the one hand and the acceptance of a broader range of options and techniques for conservation on the other.

- conservation of ecosystems or agroecosystems
- conservation of target species *within* ecosystems/agroecosystems
- conservation of semi-domesticates or pre-domesticates in home gardens or traditional farming systems
- on-farm conservation of landraces

Figure 2. Types of *in situ* conservation.

In situ conservation involves a triple set of dynamics (Fig. 3): (1) that of the environmental factors (climatic, edaphic, biotic) that affect the ecosystem itself and today involves a new factor – that of climate change; (2) that of the ecosystem itself which may show considerable change over short periods of time; and (3) that of the populations of the species that make up the ecosystem that may fluctuate considerably in size, distribution, genetic and composition even from one year to another. This dynamism has considerable implications for conservation since it can lead to considerable species turnover and even local loss (especially of rare species) even in areas that have been set aside for conservation.

- Environmental factors: Climatic, edaphic, biotic. Global change (population, climatic)
- Ecosystem dynamics
- Component species' dynamics

Figure 3. Triple dynamics of *in situ* conservation.

Biotechnology and the conservation of wild species and landraces

The current pattern in the use of wild species in plant breeding is to detect and transfer genes, particularly resistant genes (Williams 1990). Wild species are not being used for the transfer of suites of genes i.e. major components of their genotypes, since although it would be desirable to transfer adaptive complexes to cultivated material, there are major difficulties in doing so. Biotechnological techniques make possible the transfer of genetic material from unrelated species, so it could be argued that the need for collecting and conserving wild relatives of crops is diminished. On the other hand, suitable alleles or desirable traits such as hardiness, salt resistance, disease resistance, etc., are more likely to be found in crop relatives than in unrelated species and biotechnology will facilitate the transfer of genes from species that belong to the secondary and tertiary gene pools of crops, something that is technically difficult by conventional breeding techniques.

The role of biotechnology, in the broad sense, in the conservation of wild species in a non-agricultural context is mainly the application of techniques such as micropropagation for the maintenance, reinforcement, recovery and reintroduction of endangered species, and in assessing the genetic variation and patterns of geneflow in wild populations. There is an extensive literature on this field (e.g. Falk and Holsinger 1991, Valdés et al. 1997) and it will not be reviewed here.

Conservation strategies for landraces

According to Frankel and Brown (1995) the present and future role of landraces in modern agroecosystems is to serve as sources of genetic materials for plant improvement.

The conservation of landraces poses both technical and conceptual problems. The primary approach of conserving samples (accessions) of landraces as seed in gene banks or in field genebanks is by no means straightforward – should one collect alleles or genotypes for example? Whatever is conserved is not in any real sense the landrace itself because of the impossibility of conserving in such a way the dynamic socio-cultural genecological adaptations that are their very essence. This has led to attempts to conserve landraces on-farm so as to maintain these relationships. Such an approach was earlier rejected as being impracticable. Once landraces have been displaced by improved cultivars, as part of change in the agronomic and social situation, such an attempt to retain 'primitive farms' in a radically changed social and technological climate would, except in the short term, be self-defeating (Frankel and Soulé 1981, Frankel and Brown 1995).

More recently, on-farm conservation has gained more favour. For example, on-farm conservation is one of the priority activities in the Global Plan of Action: the long-term objective being 'To better understand and improve the effectiveness of on-farm conservation, management, improvement and use of plant genetic resources for food and agriculture'. And strengthening the scientific basis of *in situ* conservation of agricultural biodiversity on-farm, the subject of an IPGRI initiative (Jarvis and Hodgkin 1998), involves describing and understanding all of the factors which influence farmer practices and decision-making with relation to the maintenance of genetic diversity.

Conservation of landraces *in situ* is already occurring in both developing and developed countries. In Ethiopia, for example, the Seeds of Survival Program/Ethiopia is a participatory, dynamic, farmer-based approach to the conservation of landraces and their enhancement and utilization through on-farm multiplication and distribution (Worede et al. 1999), and an on-farm landrace conservation project, 'A Dynamic Farmer-Based Approach to the Conservation of Ethiopia's Plant Genetic Resources' funded by the GEF and the Biodiversity Institute of Ethiopia (Asfaw 1999), and there are other examples in Africa, South West Asia, Asia, Latin America.

But again, in line with our interpretation of conservation as management for change, then we have to consider carefully what are the aims of on-farm conservation and what kind if diversity it is wished to preserve – characteristic phenotypes, culinary properties, features associated with environmental adaptation or what? As Louette (1999) points out, there is no single answer to these questions. The classic idea of conserving the genetic diversity of a landrace is by 'freezing the genetic landscape', in the words of Iltis (1974), by isolating it in time and space, while at the same time maintaining the socio-economic-cultural context in which it occurs, is being questioned. Several authors have suggested replacing it by a dynamic model that recognizes that agroecosystems are themselves dynamic and that local varieties exist in a system that extends beyond a single place (Louette 1999). This approach is consistent with the view of Frankel and Soulé (1981) who point out that the genetic diversity of landraces has two dimensions: between sites and populations, and within sites and populations. Such an approach considers that what is important to preserve is not the genetic material itself but the processes that create and preserve genetic diversity. If it is specific alleles that one wishes to preserve, then *ex situ* conservation of accessions of the landrace in a seed bank would be a more appropriate strategy.

Similarly, the idea that *in situ* conservation is always meaningful concept in developing/traditional agriculture is being questioned (Gonzales 1999). Cromwell and van Oosterhout (1999) state that evidence is gathering to suggest that farmers manage rather than conserve (or preserve) on-farm diversity: their experience in Zimbabwe is that 'farmers do not preserve a static portfolio of crops and crop varieties on their farms, nor do they prevent introgression from neighbor's fields, field margins, fallow fields, or other areas where wild crop relatives grow, but rather they import and discard diversity in a dynamic fashion, according to their needs in any given period of time'. They further suggest that 'most "on-farm conservation" projects that succeed in motivating farmers to preserve individual crops and crop varieties, are reliant on compensating farmers, usually through payment. This can result in reducing the dynamism and flexibility in the farming system, because farmers do not usually view on-farm crop diversity in a static way, but as part of their constant struggle to achieve sustainable livelihoods". The conclusion that one can draw is that it may not be possible to achieve long-term conservation of individual crops and crop varieties using farmers' existing management strategies, although it may be possible to support farmers in maintaining a dynamic portfolio of on-farm crop diversity.

Threats posed by biotechnology to agrobiodiversity conservation

Landraces/local cultivars have often been the donors in breeding programme of many useful traits and as Swaminathan (2000) has put it, under small farm conditions, such as those that prevail in India, every farm is a genetic garden containing numerous varieties of each of the annual and perennial crops. There is a real risk that GM foods will lead to the replacement of numerous local cultivars or landraces by one or two genetically engineered varieties, thus leading to an acceleration of the process of genetic erosion that is already occurring through the introduction of conventional high yielding cultivars. Whatever the advantages of GM crops and foods, we must not dismiss this narrowing of the genetic base of food security by countering that biotechnology actually helps protects biodiversity and natural ecosystems in other ways, such as through as increasing yields and thus reducing the pressure on the need to cultivate new land. As Swaminathan (2000) comments, we need to use every means at our disposal to enlarge the food basket, not reduce it.

Organizations for the conservation of heritage varieties

Many countries have created organizations for the maintenance of ancient farm, fruit tree or garden varieties in seed banks, field genebanks or on-farm. These may be both government-supported or privately financed (including by NGOs). In the European Union, legislation that affects the sale of crop seeds is more restrictive than anywhere else in the world (Cherfas 2000). Although it was introduced to prevent abuses such as the adulteration of seed or repackaging under other names, it means in effect that unless a cultivar is registered in at least one of the EU Member States, it may not be marketed. Thus ancient or 'heirloom' varieties or cultivars that have simply fallen out of favour cannot be sold if not properly registered. Although there are similar laws in countries outside Europe to protect cultivars, they do allow anyone to sell unregistered seeds under normal consumer protection alone.

Organizations for the conservation of such material are therefore of considerable importance for the maintenance of diversity. A valuable survey of the methodological issues involved is given by Marchenay (1986a) and a practical guide by the same author (Marchenay, 1986b). Extensive information on seedsaving and seedsavers' resources may be obtained from the website http://homepage.tinet.ie/~merlyn/seedsaving.html.

Examples of such organizations include Switzerland's 'Pro Specie Rara' Foundation (Bossard 1999) which includes in its aims:

- Collection and sharing of planting material of delisted old cultivars and landraces.
- Involvement of a great diversity of people in the conservation of biodiversity.
- Sustaining the operations through financial inputs and inputs in kind.
- Focus not only on genetics but also on cultivation.
- Focus not only on conservation but on development.

Other European examples include:

- 'Arche Noah' in Austria with a large collection of vegetable, flower and cereal varieties
- The UK government-supported Heritage Seed Library and the Henry Doubleday Research Association, with its collection of traditional vegetable varieties
- The Biological Farming Association 'Elkana' in Georgia and the Dika, an NGO aimed at the *in situ*/on-farm conservation of Georgia's agrobiodiversity.
- The French Conservatoires botaniques nationaux.
- The French Conservatoire national des plantes à parfum, médicinales et industrielles.
- The French Conservatoire des collections végétales spécialisées.
- The French Association française pour la conservation des espèces végétales.
- The Nordic Gene Bank clonal collection of old fruit varieties.
- The German Bauerngärten (traditional peasant gardens) that conserve old local fruit varieties.

In the USA, comparable organizations include the Seed Savers Exchange, Native Seed Search and Southern Seed Legacy (Rhoades and Nazarea 1998).

There is, of course, significant difference between maintaining samples of heritage varieties in an orchard in Switzerland, for example, by organizations such as Pro Specie Rara, and maintaining primitive cultivars on-farm in a developing country. Both play an important role, however, as repositories of genetic diversity of interest to agriculture and horticulture.

Of course, the reasons for conservation of landraces and heritage varieties are various – social, cultural, practical – as well as to ensure the availability of genetic diversity for future breeding programmes. As Juma (1989) notes, the narrowing of choice imposed by the narrowing genetic basis of major commercial food crops does not reflect the preferences of consumers who often opt for greater diversity in their food sources when able to exercise a choice. This applies not only to developing countries but is one of the main reasons for the conservation of heritage varieties of fruits, flowers and vegetables in the developed world. Thus it is noteworthy that some of the main potato cultivars available and widely sold in British supermarkets are essentially the same as those of fifty to a hundred or more years ago, such as 'Kerr's Pink, 'Golden Wonder', 'Majestic' and 'King Edward', varieties that are still favoured by the consumer because of their flavour and cooking properties, despite their greater susceptibility to disease etc. than more recent cultivars.

Climate change

An important, and some would say, limiting factor that should colour approaches to the conservation of genetic diversity for the needs of future crop production is climate change. Although there are still major divergences concerning the exact nature and pattern of such change, few if any responsible scientists today would deny that change is happening and the uncertainty as to the detail simply adds to the difficulties of coping with it.

Climate change is one of the major factors that contribute to the dynamism of both ecosystems and species' populations that have to be taken into account when planning a biodiversity conservation strategy. There is already considerable evidence to suggest that the rate of climatic change in some areas will be considerably faster than the ability if species' populations to migrate or adapt, and that as a consequence new assemblages of species will be formed as a result of differential migration of individual species.

Similarly, when planning a conservation strategy for components of agrobiodiversity, such as landraces, it is clear that we must consider a number questions such as, (1) what effects will climate change have on current agroecosystems and crops? (2) what will be the effects on the habitats of the wider wild genepools of crop? (3) which landraces should be collected and maintained and for what kinds of characteristic? (Williams 1990).

One of the effects of climate change will be to put at risk landraces of crops that remain in remote areas: unless steps are taken to collect and conserve genetic samples of these (either *ex situ* or, less likely, on-farm), as a matter of urgency, they are likely to disappear soon. And it is areas such as arid, mountain or other stress environments where landraces tend to persist today, whether by accident or design, and these are often precisely the kind of landraces where desirable traits in demand today may be found. Modern improved cultivars have been readily introduced and accepted in areas that are relatively homogeneous whereas in marginal areas where the conditions are environmentally, socially, economically and culturally diverse, landraces have been retained (Brush 1999).

Fragmentation

An important factor that has to be taken into account in the design of conservation strategies is the phenomenon of fragmentation. Fragmentation or patchiness of vegetation is an almost universal phenomenon (Heywood 1999c) and habitat fragmentation has been described as a central concern in conservation biology (Harrison 1994). As Margules (1987) observes, 'Given the existing patchiness of much habitat throughout the world, research should be concentrated on the management problems of small patches, ... rather than with an unhealthy academic debate on the size and shape arrangements of reserve networks. Although the debate is entrenched in the ecological literature, so rarely is there an opportunity to plan reserves prior to habitat fragmentation, it is not really a major conservation issue.' Saunders et al. (1991) consider that while there has been much focus on the design of nature reserves, we are usually too late to do anything but maintain the remnants left following fragmentation. Ecosystem management has to concern itself with both the fragmented ecosystems and the populations of the species that are found in them.

Of course, fragmentation is not a new phenomenon – it is a process that has been going on since the beginnings of agriculture. What is alarming is the global scale of such fragmentation, especially in the past few decades, and the underlying socio-economic factors that encourage it. The consequence is that there are hardly any regions of the world where substantial tracts of intact 'undisturbed' vegetation of any kind can be found and the few large blocks of more or less intact vegetation that do remain, such as the large, unbroken stands of lowland humid forest vegetation in the Amazon basin, the Congo basin and in parts of south-east Asia, seem likely to follow the same path as a matter of time. Natural vegetation fragments are now becoming an almost universal component of our landscapes and as Lash (1996) has graphically put it, are 'Storm-battered islands in a sea of human settlement'.

The fragmentation of vegetation and the conversion of much wild land to other uses, has led to proposals for the adoption a broad landscape or bioregional approach in which all aspects of the landscape are taken into consideration – natural, semi-natural, heavily modified including agricultural, industrial, and urban, protected and unprotected. When it comes to management, agricultural ecosystems should not be regarded as self-contained units but viewed in the context of large biotically viable regions.

As Miller (1996), in a review of the bioregional approach, points out, 'in landscapes where protected areas have not been established, key genetic, taxonomic, and ecological elements of diversity that once may have been found in wildlands, or extensive farm or forest operations, are now relegated to isolated patches in intensively managed farms, pastures, timber-harvesting sites, and suburban, urban, and industrial areas'.

Conclusions

A broad perspective of agrobiodiversity is proposed and the need for such a wide approach in considering the significance of biotechnology in agriculture is stressed. Conservation strategies for agrobiodiversity likewise need to be broadly based and complementary, paralleling similar developments in the conservation of wild biodiversity. Attention is drawn to the dynamics of biodiversity and agrobiodiversity and the need to take this into account when planning management regimes.

A broadly based approach to genetic conservation of agrobiodiversity is also proposed and attention drawn to the large range of species, wild, semi-domesticated and cultivated that are used in traditional farm systems and contribute to farm household economies. Although there is considerable potential for biotechnology to contribute to improved yields and reduced risks for poor farmers, as well as more plentiful, affordable and nutritious food for poor consumers, as Pinstrup-Andersen and Cohen (2000), observe, it is unlikely that biotechnological techniques will be applied to the majority of the species involved so that reliance on traditional methods of selection and breeding will continue to be needed to enhance productivity. As they say, 'It is ... critical that biotechnology be viewed as one part of a comprehensive sustainable poverty alleviation strategy, not a technological quick fix for world hunger.

References

Altieri MA (1998) The agroecological dimensions of biodiversity in traditional farming systems. In: Posey DA (ed.): *Cultural and Spiritual Values of Biodiversity* 291–297. For UNEP, Intermediate Technology Publications, London

Angermeier PI (1994) Does biodiversity include artificial diversity? *Conservation Biology* 8: 600–602

Asfaw Z (1999) The barleys of Ethiopia. In: Brush, S.B. (ed.): *Genes in the Field. On-farm conservation of crop diversity*, 77–107. IPGRI, IDRC & Lewis Publishers, Boca Raton

Ayad WG, Hodgkin T, Jaradat A, Rao VR (eds.): *Molecular Genetic Techniques for Plant Genetic Resources. Report of an IPGRI Workshop 9–11 October 1995, Rome Italy*. IPGRI, Rome

Bossard M (1999) Short introduction to the conservation work of the Swiss 'Pro Specie Rara'. In: Gass T, Frese L, Begemann F, Lipman E (eds.), *Implementation of the Global Plan of Action in Europe – Conservation and Sustainable Utilization of Plant Genetic Resources for Food and Agriculture. Proceedings of the European Symposium, 30 June–3 July 1988, Braunschweig, Germany*. International Plant Genetic Resources Institute, Rome, 1999, 63–65

Brush SB (1999) The issues of *in situ* conservation of crop genetic resources. In: Brush SB (ed.): *Genes in the Field. On farm conservation of crop diversity*, 3–25. Publishers, New York, IDRC, Ottawa & IPGRI, Rome

Busch L, Lacy WB, Burkhardt J (1989) Culture and care: ethical and policy dimensions of germplasm conservation. In: Knutson L, Stoner AK (eds.): *Biotic Diversity and Germplasm Conservation, Global Imperatives,* 43–62, Kluwer Academic, Dordrecht

Changtragoon S, Szmidt AE, Wang X-R (1987) The use of molecular markers in the study of genetic diversity in ratten: preliminary results. In: Ayad WG, Hodgkin T, Jaradat A, Rao VR (eds.): *Molecular Genetic Techniques for Plant Genetic Resources. Report of an IPGRI Workshop 9–11 October 1995, Rome Italy*. IPGRI, Rome

Cherfas J (2000) Legalise diversity. *The Garden* 125:212–213
Cohen JI (1992) Conservation and Use of Agro-ecological Diversity. ACTS Biopolicy International Series No. 3, Acts Press, Nairobi
Conway GR (1987) The properties of agroecosystems. *Agricultural Systems* 24: 95–117
Conway GR (1997) *The Doubly Green Revolution. Food for all in the 21st century.* Comstock Publishing Associates, Cornell Universty Press, Ithaca, New York
Cromwell E, van Oosterhout S (1999) On-farm conservation of crop diversity: policy and institutional lessons from Zimbabwe. In: Brush SB (ed.): *Genes in the Field. On-farm conservation of crop diversity*, 217–238. IPGRI, IDRC & Lewis Publishers, Boca Raton
Davis PH, Heywood VH (eds.) (1963) *Principles of Angiosperm Taxonomy.* Oliver & Boyd, Edinburgh
Dudley N (ed.) (1992) *Forests in Trouble: A review of the status of temperate forests worldwide.* Worldwide Fund for Nature, Gland
Ehrlich PR, Murphy DD (1987) Conservation lessons from long-term studies of checkerspot butterflies. *Conservation Biology* 1: 122–131
Falk DA (1990) Integrated strategies for conserving plant genetic diversity. *Ann. Mo. Bot. Gard* 77: 38–47
FAO (1996) *Global Plan of Action for the Conservation and Sustainable Use of Plant Genetic Resources for Food and Agriculture.* FAO, Rome
Frankel OH, Brown AHD, Burdon JJ (eds.) (1995) *The Conservation of Plant Biodiversity.* Cambridge University Press, Cambridge
Frankel OH, Soulé ME (eds.) (1981) *Conservation and Evolution.* Cambridge University Press, London
Falk DA, Holsinger KE (eds.) (1991) *Genetics and Conservation of Rare Plants.* Oxford University Press, New York
Fowler C, Hawtin GC, Hodgkin T (1999) Foreword. In: Brush SB (ed.): *Genes in the Field. On farm conservation of crop diversity.* Publishers, New York, IDRC, Ottawa & IPGRI, Rome
Gonzales TA (1999) The cultures of the seed in the Peruvian Andes. In: Brush SB (ed.): *Genes in the Field. On-farm conservation of crop diversity*, 193–216. IPGRI, IDRC & Lewis Publishers, Boca Raton
Gregor JG, Horne FR (1935) Growth forms: genecology and its agricultural significance. *Agric. Progress* 12: 89–98
Halan JR (1975) Our vanishing genetic resources. *Science* 188:618–21
Harrison S (1994) Metapopulations and conservation. In: Edwards PJ, May RM, Webb N R (eds.): *Large-scale Ecology and Conservation Biology* 111–128
Hawkes JG (1983) *The Diversity of Crop Plants.* Harvard University Press, Cambridge MA
Hawkes JG (1990) N.I. Vavilov – the man and his work. *Biol. J. Linn.Soc.* 39: 3–6
Heywood VH (1993) Broadening the basis of plant resource conservation. In: Gustafson JP, Appels R, Raven P (eds.): *Gene Conservation and Exploitation*, 1–13. Plenum, Press, New York
Heywood VH (ed.) (1995) *Global Biodiversity Assessment.* UNEP. Cambridge University Press
Heywood (1999a) Trends in agricultural biodiversity. In: Janick J (ed.): *Perspectives on New Crops and New Uses*, ASHS Press, Alexandria, Va, 1999, 2-14
Heywood VH (1999b) *Use and Potential of Wild Plants in Farm Households.* FAO, Rome
Heywood VH (1999c) Is the conservation of vegetation fragments and their biodiversity worth the effort? In: Maltby E, Holdgate M, Acreman M, Weir AG (eds.): *Ecosystem Management: Questions for science and society.* Royal Holloway Institute for Environmental Research, Royal Holloway, University of London, UK, 1998, 65-76
Holden J, Peacock J, Williams T (eds.) (1993) *Genes, Crops and the Environment.* Cambridge University Press Cambridge
Iltis HH (1974) Freezing the genetic landscape: the preservation of diversity in cultivated plants as an urgent social responsibility of plant geneticist and plant taxonomist. *Maize Genetics Cooperation News Letter* 48: 199–200
Jansen PCM, Lemmens RHMJ, Oyen LPA, Siemonsa JS, Stavast FM, van Valkenburg JLCH (eds.): *Plant Resources of South-East Asia. Basic list of species and commodity groupings.* Final version. PROSEA Foundation, Bogor
Jarvis DI, Hodgkin T (eds.) (1998) *Strengthening the scientific basis of in situ conservation of agricultural biodiversity on-far. Options for data collecting and analysis. Proceedings of a workshop to develop tools and procedures for* in situ *conservation on-farm, 25–29 August 1997, Rome, Italy.* IPGRI, Rome
Juma C (ed.) (1989) *The Gene Hunters: Biotechnology and the Scramble for Seeds.* Princeton University Press
Lash J (1996) Preface. In: Miller KR (ed.): *Balancing the Scales: Guidelines for increasing biodiversity's chances through bioregional management.* World Resources Institute, Washington DC
Lawton JH, Jones CG (1993) Linking species and ecosystem perspectives. *Trends in Ecology and Evolution* 8: 311–313

Louette D (1999) Traditional management of seed and genetic diversity: what is a landrace? In: Brush SB (ed.): *Genes in the Field. On-farm conservation of crop diversity*, 109–142. IPGRI, IDRC & Lewis Publishers, Boca Raton

Marchenay P (ed.) (1986a.) *A la Recherche des Variétés Locales de Plantes Cultivées*. PAGE-PACA, Conservatoire Botanique de Porquerolles, Hyères

Marchenay P (ed.) (1986b) *Prospection et Collecte des Variétés Locales de Plantes Cultivées.*Guide Pratique. PAGE-PACA, Conservatoire Botanique de Porquerolles, Hyères

Margules CR (1987) Single large or several small reserves? In: Saunders DA, Arnold GW, Burbidge AA, Hopkins AJM (eds.): *Nature Conservation: The Role of Remnants of Native Vegetation*. 361. Surrey Beatty and Sons Pty in association with CSIRO and CALM

McCalla AF, Brown LR (2000) Feeding the world in the next millennium: a question of science? In: Persley GJ, Lantin MM (eds.): *Agricultural Biotechnology and the Poor*. 32–36. CGIAR, Washington

Miller KR (1996) Balancing the Scales: Guidelines for increasing biodiversity's chances through bioregional management. World Resources Institute, Washington DC

Persley GJ, Lantin MM (eds.) (2000) *Agricultural Biotechnology and the Poor.* CGIAR, Washington

Pinstrup-Andersen P, Cohen MJ (2000) Modern biotechnology for food and agriculture: risks and opportunities for the poor. In: Persley GJ, Lantin MM (eds.), *Agricultural Biotechnology and the Poor*. 159–169. CGIAR, Washington

Plucknett DL, Smith NJH, Williams JT, Anishetty M (1987) *Gene Banks and the World's Food.* Princeton University Press, N.J

Prescott-Allen R, Prescott-Allen C (1990) How many plants feed the world? *Conservation Biology* 4: 365–374.

Raven PH (1998) Commentary in: Planet of the plants. *World Conservation* 2/98: 22–23

Rhoades RE, Nazarea VD (1998) Local management of diversity in traditional agroecosystems. In: Collins WW, Qualset CO (eds.): *Biodiversty in Agroecosystems*, 215–236

Saunders DA, Hobbs RJ, Margules CR (1991) Biological consequences of ecosystem fragmentation: a review. *Conservation Biology* 5: 18–32

Seyani J (1998) Commentary in: Planet of the plants. *World Conservation* 2/98: 22–23

Soulé ME, Mills LS (1992) Conservation genetics and conservation biology: a troubled marriage. In: Sandlund OT, Hindar K, Brown AHD (eds.): *Conservation of Biodiversity for Sustainable Development*, 55–69. Scandinavian Univetsity Press, Oslo

Swaminathan MS (2000) Genetic engineering and food security: ecological and livelihood issues. In: Persley GJ, Lantin MM (eds.): *Agricultural Biotechnology and the Poor*. 37–42. CGIAR, Washington

Turner IM (1996) Species loss in fragments of tropical rain forest: a review of the evidence. *J. Appl. Ecol.* 33:200-219

Thrupp LA (ed.) (1996) *New Partnerships for Sustainable Agriculture*. World Resources Institute, Washington.

Valdés B, Heywood VH, Raimondo FM, Zohary D (eds.) (1997) *Proceedings of the Workshops on the Conservation of the Wild Relatives of European Cultivated Plants.* (*Bocconea* 7). Azienda Foreste Demaniali della Regione Siciliana, Palermo

Vandermeer J, Perfecto I (1997) The agroecosystem: a need for the biologist's lens. *Conservation Biology* 11: 591–592

WARDA (West Africa Rice Development Association) (1999) *The Spark that Lit a Flame: Participatory varietal selection.* WARDA, Bouaké, Côte d'Ivoire

Williams JT (1990) The implications of climatic changes for plant genetic conservation strategies. In: M Jackson, Ford-Lloyd BV, Parry ML (eds.): *Climate Change and Plant Genetic Resources*, 167–178

Worede M, Tesemma T, Feyissa R (1999) Keeping diversity alive: an Ethiopian perspective. In: SB Brush (ed): *Genes in the Field. On-farm conservation of crop diversity*, 143–161. IPGRI, IDRC & Lewis Publishers, Boca Raton

Session 4: Conclusions

© 2003 Birkhäuser Verlag Basel/Switzerland

Discussion: The way ahead

Edited from tape recording by Klaus Ammann, Botanical garden, University of Bern, Altenbergrain 21, CH- 3013 Bern

Klaus Ammann: Introductory remarks.
I am very much looking forward to this debate. Let's set it up as a discursive process, since we have so many specialists from various fields of this important debate about biotechnology, have a debate on solutions for the major problems. We should be more modest and choose a step wise, argumentative approach and my role I see as a moderator, which acts more as a midwife than as an opinion leader trying to influence everybody with my ideas. We need lots of grass root approaches, under the spirit of the symmetry of ignorance. If we succeed to listen to each other in an often advocated collaborated learning atmosphere, I will be happy.

Richard Braun
In such a high power meeting it would be nice to interlink to a manuscript in statu nascenti about biotechnology and biotechnology, written for the European Federation of Biotechnology. I will write it with the help of Klaus Ammann, and we both would love to get a feedback from you as participants.

Alan Gray
During the last two days one aspect was repeatedly stated: The whole biotech – biodiversity issue is one of context and perspective, and one which should be set within the context of traditional agriculture for instance. We would miss an important opportunity, if we did not place the risks from biotechnology in perspective, incorporating the risks both from GMO's and from traditional agriculture, perhaps setting these in a historical context. The public would not understand if we were to miss this opportunity.

One other point: It will be difficult to come up with a consensus definition for biodiversity. Think about the difficulties we meet in defining even a basic concept such as "species".

Richard Braun
Definitions have been attempted in such basic terms hundreds of times; I don't think there is time to go into such details.

Klaus Ammann
Vernon Heywood gave a nice and very broad minded definition of biodiversity. Think about the very questionable definition of GMO's: We all know as biologists that this is a totally misleading expression, since basically we all eat genetically modified crops since centuries, and – after all, we are all genetically modified ourselves. Therefore, instead of going into definition problems, we should concentrate on project oriented debates, where we could come up with visions for solutions.

Anatole Krattiger
First a question: what about the structure of the debate, Klaus?

Klaus Ammann
The task is to throw in some ideas which later can be structured.

Alan Gray
Focus on the question as to how modern agriculture may be put into the perspective of traditional agriculture, concentrating on the fitness of all novel genes in relation to field releases. Is it really justified to single out the products of genetic engineering from those of traditional breeding?

Jill Turner
I think we need more than a briefing paper on the subject. We should invest time into more project oriented outcomes.

Richard Braun
Let's depart from this briefing paper issue. Anatole has lots of topics to suggest, many other people here as well.

Anatole Krattiger
Actually it is at this stage more a process and future oriented proposal on the structure of our debate: Suggestions could be bundled into a matrix with 4-5 columns. Various ideas in the first column. Columns on biodiversity, productivity and put into the perspective all proposals on whether they are reducing or enhancing biotechnology and productivity – or are neutral and enhance biodiversity alone - and how to mitigate negative impacts. Then we could bundle our thoughts to five lines: 1) Policy, 2) Research, 3) Public–private partnerships, 4) Institutional issues, 5) Implementation priorities.

Hugo Mösch
Anatole Krattiger's proposal is so perfect that we all might go home.

Jill Turner
I would like to make an argument for the inclusion of social issues in any project which develops. We need to not only enhance communication as I emphasised yesterday. It is not only biodiversity that is a focus this century but certainly human genetics will also be given historical focus. We must bring this altogether. Vernon Heywood made an excellent remark: "Fragmentation is the major issue of modernity."

Richard Braun
I agree with Jill Turner and also Vernon Heywood about the social issues, we would be wrong not including them in all projects. We are still in the preamble phase in filling in Anatole Krattiger's table.

Paul Egger
If we look at the issues on biodiversity and biotechnology, we also need to assess and take into account research capacity. We will have a concentration on some 30 crops, this will have a strong influence on biodiversity, and we need therefore to look at public-private linkages in research and the markets.

Vernon Heywood
What baseline data do we need when judging the risks about the agroecosystems we are going to place GMO's into? We need a list of important data for environmental impact assessment.

Richard Braun
A list of research priorities would also be very useful.

Jill Turner
We should look at a truly international perspective, and we should not only follow the research and regulatory priorities of USDA. We also should address hunger.

Brian Johnson
The UK regulatory system is already working on a generic list of questions that need to be answered if we want to address the risks of biotechnology related to biodiversity. The draft list contains generic questions that could be also be applied to tropical countries. The release of any new crops is a question that is particularly acute in tropical countries. Seen that way we could apply such a list of generic questions not only in East Anglia in the UK, but also in Senegal. We should have a good framework within the next few months.

One set of questions relates to impacts of management systems used at the moment; what effect those currently have on biodiversity.

Sandy Thomas
We haven't really gone into the issues of access. Germ plasm will fall more and more into the control of the life science companies. We must address such issues, since they will be key area when it comes to questions of delivery of modern technologies. We do not have up to know a commitment of the companies to do something for the developing world.

Jill Turner
Another issue: If we are to preserve landraces, the question will be what qualities of landraces do we want to preserve? Should we focus on local and traditional agriculture or on the people themselves: Do we want to preserve just a few important alleles of genes or do we want to preserve land races as a whole?

Maria Ruohonen-Lehto
My priorities are local people and their problems of hunger. Should we really concentrate on agreeing on general principles? Implementation of modern breeding is a major issue here.

Klaus Ammann
I am sure that all the problems we discuss here have a communication, social and ethical component.

We should soon stop this kind of unfocussed brainstorming. We should not try to create a huge UNEP programme, otherwise we will end up in various kinds of priorities which will be hard to bundle.

My proposal would be that we should concentrate on a baseline data list on risk and benefit data to be presented and also combine it with a list of characters, which should be implemented by any kind of plant breeding. It would be excellent if plant breeders of the public and private sector could get a feedback from regulators and scientists actively involved in risk assessment research.

Tom Nickson
I sympathise with much what I hear here. But I see problems: Regulatory frameworks are really not harmonized, and the risk assessment framework that exists today is still being developed: we have problem formulation, we have risk analysis which is at a data accumulation stage, risk characterisation followed then by risk management, which is not yet in a satisfactory stage in development. And, we need crop and even trait specific data, generalizations cannot be accepted.

Sandy Thomas
I am not sure whether we should leave the issue on public and private funding. What is the route of innovation for new cultivars? It will be influenced by funding for sure. This should be addressed by this group, even if it will not be in the main thrust of our activity here.

Brian Johnson
I want to come back to the important remarks of Tom Nickson: the point about risk assessment. Many of the points made in this workshop are not addressed in the present day risk assessment framework. The risk assessment process is concentrating on direct effects, the framework needs to be extended to indirect effects.

Tom Nickson
I agree with Brian Johnson about focus being on direct effects, however the problem formulation is intended to include indirect effects.

Brian Johnson
The regulatory system should include foresight and feedback to the industry itself. Klaus Ammann's question is a very important question: How to design new cultivars, should we develop new concepts, which are better than those we developed 10 years ago.

Jill Turner
We need to distinguish between the risks of biotechnology on biodiversity or on agriculture in general. We need something like a baseline and to take into account a risk-benefit perspective. If we can focus on the risks related to biotechnology we could then better develop a catalogue of what modern crop characteristics should look like.

Hugo Mösch
Maybe we should also consider the concepts of Greenpeace on GMO's: 1) Exchange of genes between species is still very poorly understood. 2) There are no adequate ecological concepts to predict the interaction of GMO's with the environment.

Klaus Ammann
You can always say this because those concepts will never really exist, regionalisation and complexity of the environment will always contradict simple concepts. And if we would act according to this principle, there would be no modern agriculture at all.

Barry Palevitz
This is again the point I always make: We must make clear the context to traditional agriculture, and if you want to talk to the big public, you must make this point very clear, otherwise you miss an important opportunity.

Hans Herren
The idea was to have specific projects, let's go back to "grasses". Let's start with a concrete project on grasses.

Klaus Ammann
We should now work on the table in order to put some logic structure into our discussion.

Tom Nickson
I am a little lost here about the regulatory framework, which is not harmonized. As a scientist I am more interested in risk assessment research and its methodology.

Brian Johnson
I have some problems with the word risk. Risk assessment is essentially concerned with safety, but does not take into account many indirect impacts of introducing new traits. Maybe we should use the phrase "sustainability assessment", and that would bring us automatically to a more balanced view of risks and benefits. A crop variety is not simply a genetic unit related to plowing and yield, but it enables a whole system of human agricultural practice.

Tom Nickson
But risk assessment is derived from regulatory frameworks which are taking into account broader concepts such as sustainability.

Klaus Ammann
A proposal: Let's divide up in three discursive groups with the following topics: 1) Research priorities, 2) safety sustainability and 3) public-private partnership issues.

Anatole Krattiger
We also should have a category on what we have learned in those two days. And more concretely: What, if you would get 5 million $, what would you do as a priority? This would add specificity to the discussions and prevent us from wandering off in tangents.

We could set up a few groups, say 3 to 6 groups each with their own chairs to elaborate on how they would spend the 5 million $, and then report back to the plenary after one hour. This would allow us to fill out the matrix we where starting to construct.

Barry Palevitz
In any successful strategy you have to consider the public, it is out there and needs to be addressed. But who is the public; it's a heterogeneous mass of people with totally diverging interests.

Anatole Krattiger
The target group depends on the topic and the group you are working in.

Sandy Thomas
We should avoid to just talk to ourselves – and this despite the complexity of the issues we work on in this meeting. I really miss the complexity of the problems when we talk to the public.

Anatole Krattiger
Let's divide up into small subgroups for further developing of the topics:
- Chair for research: Klaus Ammann, Vernon Heywood
- Chair for Safety and Sustainability Brian Johnson, Tom Nickson
- Chair Partnerships Paul Egger, Sandy Thomas

5 min. presentation Klaus Ammann on research priorities:
Biodiversity and consensus is just a contradiction in itself.
1) Dynamics and history of agricultural systems: We should look more deeply into the dynamics of agricultural systems. We should reinforce agro-archaeology for the study of ancient seeds, follow up more closely weed development. Basically we should make use of other "time machines" such as herbaria, pollen archives in peat and many other soil types, seed archives of the past and also have a closer look at the remainings of plants and animals.
2) Calibrated biomonitoring: We should correlate a selected set of organisms to soil fertility, soil pollution with heavy metals. It should be possible to cut through with selected biomonitoring organisms the gordic knot of the overwhelming complexity of the ecological system, so that for the future we can develop a regionalised set of biomonitoring systems to know in a comparative, semiquantitative way what is going on in the fields. This has been done with success in the research field of biomonitoring air pollution with lichens.
3) Precision biotechnology: It would be beneficial to develop new concepts of agricultural practice: As an example we propose to design new patterns of resistance, which are closer to nature: Plants are themselves producing thousands of natural pesticides and they are unevenly distributed among the natural populations. It would be interesting to achieve this with biotechnology again, since we have lost part of this natural mixture of resistance in crops. Why not develop mixtures of a variety of resistances, not all stacked in one seed, but having those maybe 20 differently engineered resistance genes in a variety of seeds, but all with the same characters related to yield and regional ecological adaptation.
4) New containment methods to prevent unwelcome geneflow: We need to develop apomixis (which is occurring abundantly in nature and would be a perfect biological containment method, since apomictic plants either do not produce pollen grains at all or only unviable ones). Also we should further develop transplastomic strategies, although knowing, that it is not a perfect containment system, since there are genera such as alfalfa, which still offer the possibility of transporting plastids within their pollen grains, others just occasionally.
5) We need to develop, learn and mimic other natural systems of resistance strategies, such as systemic acquired resistance, which could be enhanced as an answer of plants attacked by pests by quickly producing natural pesticides. We are sure that we also could learn from organic farming on how to reduce the risk of pesticide attacks.

Vernon Heywood
As chair of the group I decided to abscond with the five millions to set up my own eco-religion.

Five research priorities
1) An array of baseline data on biodiversity: Comparison between the agricultural systems, adjacent impact of agriculture on natural and semi-natural systems. We would look at parameters like weed species, soils, the input of production systems, organic versus conventional, what yield in terms of proteins and calories, the efficiency of the systems etc., a comparison of an agriculture basically still producing food for the small scale farmers themselves versus an a large scale farming system where nobody on the farm eats the products directly etc.

2) Lessons from the past: What lessons we have learned from past agronomic practices, what patterns of geneflow occurred in the past using genomics, and what past impact this has had.
3) Centers of crop diversification: We want to make a very strong case in undertaking research in centers of crop diversification. Patterns of gene exchange, looking how genes have been integrated in wild relatives of crops and vice versa, look at the cultural practices related to landraces. Look at the genetic diversity of crops, also local ones.
4) Climatic change and climatic instability: We should have a closer look at climatic instability, which is not quite the same as climate change and global warming.
5) Changes in agricultural practice: Food crops in the beginning, now it is time for fuel and pharma crops, we should monitor the situation carefully.

Brian Johnson, safety and sustainability:
We first had a general discussion on safety and sustainability and public communication which got us nowhere, then we decided for a concrete project which would be topical and could cause a major agricultural shift. We decided on herbicide tolerance in rice as an appropriate model system.

GM herbicide tolerant rice could have a major impact not only on ecological, but also on social systems. A report on the relative sustainability of the new rice versus the old system would be important but could cost more than 5 million $. The new rice could enable us to grow more rice; it would enable us to initiate a transition from laborious wet rice cultivation to a more efficient dry land rice cultivation. The water could then be used for the cities. These could all be benefits. Conversely the high non-crop biodiversity of rice fields should be examined before any transition. Social aspects should also be taken into account, impacts on employment and social structures in rural areas, especially the role of women. This study would enable competent authorities and governments to make more informed choices between growing GMHT rice and conventional wetland varieties, and the choice could well depend on local conditions.

A global institution should undertake the organisation of the study, e.g. the FAO, the World Bank.

Sandy Thomas, public-private partnership:
1) Important: Advocacy role for public-private partnerships within the project would be crucial. No candidate was mentioned. The prime function: Enhance public-private partnership. Invest some money in ISAAA, which fulfils an important brokerage role. Also it would be important to have some partners and sponsors within the companies, which would enable us to negotiate about intensifying public-private partnerships. This could lead to a better access to technology for the developing world.
2) Reinforcing and strengthening public capacity. Increasing public awareness is important in developing and developed countries. Greater support for the CGIAR system would be important. Infrastructure and training should be also supported.
3) Work for the regulatory framework is important, also the implementation of the biosafety protocol would be crucial.
4) Establish a think tank for the implementation of the CBD with its important benefit sharing system. Also we would like to suggest a little study to what extent GMO's could be used to enhance biodiversity?
5) Communication is crucial. We would be wise to spend more money in communication in the sense of a good PR activity on the importance of biotechnology in plant breeding.

© 2003 Birkhäuser Verlag Basel/Switzerland

Acknowledgements

We want to sincerely thank to the following institutions for financial support:

European Federation of Biotechnology (Den Haag, Netherlands)
Max and Elsa Beer Brawand Fonds (Bern, Switzerland)

With this financial support it was possible to issue an international invitation in order to get some of the best keynote speakers within their field of specialities.

Methods for Risk Assessment of Transgenic Plants
IV. Biodiversity and Biotechnology
K. Ammann, Y. Jacot and R. Braun (eds)
© 2003 Birkhäuser Verlag Basel/Switzerland

List of contributors

Matthew Alexander
Centre for Ecology & Hydrology, CEH Dorset, Winfrith Technology Centre, Dorchester, DT2 8ZD, UK; e-mail: MAL@ceh.ac.uk

Klaus Ammann
University of Bern, Botanical Garden, Altenbergrain 21, CH – 3013 Bern, Switzerland; e-mail: klaus.ammann@ips.unibe.ch

Sir John E. Beringer
Senate House, University of Bristol, Tyndall Avenue, Bristol, BS8 1TH, UK; e-mail: J.Beringer@bristol.ac.uk

Susette Biber-Klemm
University of Basel, Faculty of Law, Maiengasse 51, CH – 4056 Basel, Switzerland; e-mail: Susette.Biber-Klemm@unibas.ch

Richard Braun
BIOLINK, P. O. Box 208, CH – 3000 Bern 11, Switzerland; e-mail: rdbraun@bluewin.ch

Ian Cooper
CEH Oxford, Mansfield Road, Oxford, OX1 3SR, UK; e-mail: jic@ceh.ac.uk

Roger Daniels
Centre for Ecology & Hydrology, CEH Dorset, Winfrith Technology Centre, Dorchester, DT2 8ZD, UK; e-mail: RED@ceh.ac.uk

Willy De Greef
IBRS, Kortrijksesteenweg 748, 9000 Gent, Belgium; e-mail: willy.degreef@ibrs.be

Hans C.M. den Nijs
Institute for Biodiversity & Ecosystem Dynamics (IBED), University of Amsterdam, Kruislaan 318, 1098 SM Amsterdam, The Netherlands; e-mail: nijs@bio.uva.nl

Mary-Lou Edwards
CEH Oxford, Mansfield Road, Oxford, OX1 3SR, UK

Alan Gray
Centre for Ecology & Hydrology, CEH Dorset, Winfrith Technology Centre, Dorchester, DT2 8ZD, UK; e-mail: ajg@ceh.ac.uk

Martin A. Girsberger
Swiss Federal Institute of Intellectual Property, Legal Services, Patent and Design Law, Division of Legal and International Affairs, Einsteinstr. 2, CH – 3003 Bern, Switzerland; e-mail: martin.girsberger@ipi.ch

Hans R. Herren
International Centre of Insect Physiology and Ecology (ICIPE), P.O. Box 30772, Nairobi, Kenya; e-mail: herren@africaonline.co.ke

Vernon Heywood
Centre for Plant Diversity and Systematics, School of Plant Sciences, The University of Reading, Reading RG6 6AS, UK; e-mail: vhheywood@btinternet.com

Brian Johnson
Biotechnology Advisory Unit, English Nature, Taunton TA1 5AA, UK; e-mail: brian.johnson@english-nature.org.uk

Anatole F. Krattiger
College of Agriculture and Life Sciences, Cornell University, 31A Warren Hall, Ithaca NY 14853, USA; e-mail: afk3@cornell.edu

Robert Lamb
Swiss Agency for Environment, Forests and Landscape, CH-3003 Bern, Switzerland; e-mail: robert.lamb@buwal.ch

Lindsay Maskell
Centre for Ecology & Hydrology, CEH Dorset, Winfrith Technology Centre, Dorchester, DT2 8ZD, UK; e-mail: LCMA@ceh.ac.uk

Thomas E. Nickson
Ecological Technology Center, Monsanto Company, 800 N. Lindbergh Blvd., St. Louis, MO 63141, USA; e-mail: thomas.nickson@monsanto.com

Beat Nobs
Swiss Agency for Environment, Forests and Landscape, CH-3003 Bern, Switzerland; e-mail: beat.nobs@buwal.admin.ch

Denise Pallett
CEH Oxford, Mansfield Road, Oxford, OX1 3SR, UK; e-mail: DWPA@ceh.ac.uk

François Pythoud
Swiss Agency for Environment, Forests and Landscape, CH-3003 Bern, Switzerland; e-mail: francois.pythoud@buwal.admin.ch

Alan Raybould
Centre for Ecology & Hydrology, CEH Dorset, Winfrith Technology Centre, Dorchester, DT2 8ZD, UK; e-mail: alan.raybould@syngenta.com

Milo Thurston
CEH Oxford, Mansfield Road, Oxford, OX1 3SR, UK

Jill Turner
The Medical Biology Center, Queen's University of Belfast, 97 Lisburn Road, Belfast, Northern Ireland BT9 7BL; e-mail: Jill.Turner@Queens-Belfast.AC.UK

© 2003 Birkhäuser Verlag Basel/Switzerland

Subject index

MIX
Papier aus verantwortungsvollen Quellen
Paper from responsible sources
FSC® C105338

If you have any concerns about our products,
you can contact us on
ProductSafety@springernature.com

In case Publisher is established outside the EU,
the EU authorized representative is:
Springer Nature Customer Service Center GmbH
Europaplatz 3, 69115 Heidelberg, Germany

Printed by Libri Plureos GmbH
in Hamburg, Germany